Collins

CONCISE REVISION COURSE

CSEC® Biology

Anne Tindale

Collins

HarperCollins Publishers Ltd
The News Building
1 London Bridge Street
London SE1 9GF

HarperCollinsPublishers
Macken House
39/40 Mayor Street Upper
Dublin 1, D01 C9W8

First edition 2016

10

ISBN 978-0-00-815787-6

www.collins.co.uk/caribbeanschools

A catalogue record for this book is available from the British Library.

Typeset by QBS
Printed in Great Britain by Martins the Printers

Author: Anne Tindale
Publisher: Elaine Higgleton
Commissioning Editor: Peter Dennis and Tom Hardy
Managing Editor: Sarah Thomas
Copy Editor: Rebecca Ramsden
Proofreader: Tim Jackson
Artwork: QBS
Cover: Kevin Robbins and Gordon MacGilp

This book contains FSC™ certified paper and other controlled sources to ensure responsible forest management.

For more information visit: www.harpercollins.co.uk/green

Acknowledgements

p4tl: 3Dalia/Shutterstock, p4tr: Wire_man/Shutterstock, p4bl: Noppharat46/Shutterstock, p4br: Bayanova Svetlana/Shutterstock, p10l: PRILL/Shutterstock, p10r: R Kristoffersen/Shutterstock, p12: Eric Isselee/Shutterstock, p12: Robert_s/Shutterstock, p12: Tatiana Belova/Shutterstock , p12: Feathercollector/Shutterstock, p12: Stephen Mcsweeny/Shutterstock, p12: Vitalii Hulai/Shutterstock, p12: Wim van Egmond/Visuals Unlimited, Inc./Getty Images, p12: Anat Chant/Shutterstock, p12: Matt9122/Shutterstock, p12: Roland Birke/Getty Images, p13: Robert L Kothenbeutel/Shutterstock, p14: Juan Gaertner/Shutterstock, p15tr: Alfredo Maiquez/Shutterstock, p15cr: Nicolas. voisin44/Shutterstock, p15cbr: Nigel Cattlin/Visuals Unlimited/Corbis, p15br: Peter Leahy/Shutterstock, p18l: Hywit Dimyadi/Shutterstock, p18r: Fabien Monteil/Shutterstock, p20: Donovan van Staden/Shutterstock, p21: Rich Carey/Shutterstock, p22: Lodimup/Shutterstock, p23: M. Shcherbyna/Shutterstock, p24: Mikadun/Shutterstock, p25l: Rich Carey/Shutterstock, p25r: Trusjom/Shutterstock, p25c: Rich Carey/Shutterstock, p34l: Dr.Jeremy Burgess/Science Photo Library, p34r: Don W Fawcett/Science Photo Library, p43: Anne Tindale, p48: Dimarion/Shutterstock, p53: Dr.M.A. Ansary/Science Photo Library, p54: Brian A Jackson/Shutterstock, p55: Mila_1989/Shutterstock, p71: Wikrom Kitsamritchai/Shutterstock, p73: Jose Luis Calvo/Shutterstock, p83: Jubal Harshaw/Shutterstock, p86l: Biodisc/Visuals Unlimited/Corbis, p86c: John Clegg/Science Photo Library, p86r: Max Topchii/Shutterstock, p89: Dourleak/Shutterstock, p105: Eranicle/Shutterstock, p126tl: Addyvanich/Shutterstock, p126tr: Southern Illinois University/Science Photo Library, p126bl: Image Point Fr/Shutterstock, p126br: Areeya_ann/Shutterstock, p151: Daimond Shutter/Shutterstock, p159t: Michael W. Tweedie/Science Photo Library, p159b: Sergey Goryachev/Shutterstock, p160: Nicky Rhodes/Shutterstock, p161: Jian Hongyan/Shutterstock

Contents

The pathway to success

About this book

This book has been written primarily as a **revision course** for students studying for the CSEC® Biology examination. The facts are presented **concisely** using a variety of formats which makes them **easy to understand** and **learn**. Key words are highlighted in **bold** type and important **definitions** which must be learnt are written in *italics* and highlighted in colour. **Annotated diagrams** and **tables** have been used wherever possible and the relationship between **structure** and **function** is continually emphasized. **Questions** to help test knowledge and understanding, and provide practice for the actual examination, are included throughout the book.

The following sections provide **valuable information** on the format of the CSEC® examination, how to revise successfully, successful examination technique, key terms used on examination papers and School-Based Assessment.

The CSEC® Biology syllabus and this book

The **CSEC® Biology syllabus** is available online at **http://cxc-store.com**. You are strongly advised to read through the syllabus carefully since it provides detailed information on the specific objectives of each topic of the course, School-Based Assessment (SBA) and the format of the CSEC® examination. Each chapter in **this book** covers a particular topic in the syllabus.

- **Chapters 1 to 4** cover topics in Section A, **Living Organisms in the Environment**
- **Chapters 5 to 15** cover topics in Section B, **Life Processes and Disease**
- **Chapters 16 and 17** cover topics in Section C, **Continuity and Variation**

At the end of each chapter, or section within a chapter, you will find a selection of **revision questions**. These questions test your **knowledge** and **understanding** of the topic covered in the chapter or section. At the end Chapters 4, 15 and 17 you will find a selection of **exam-style questions** which also test how you **apply** the knowledge you have gained and help prepare you to answer the different styles of questions that you will encounter in your CSEC® examination. You will find the answers to all these questions online at **www.collins.co.uk/caribbeanschools**.

The format of the CSEC® Biology examination

The examination consists of **two papers** and your performance is evaluated using the following three profiles:

- **Knowledge and comprehension**
- **Use of knowledge**
- **Experimental skills**

Paper 01 (1 ¼ hours)

Paper 01 consists of **60 multiple choice questions**. Each question is worth **1 mark**. Four **choices** of answer are provided for each question of which one is correct.

- Make sure you read each question **thoroughly**; some questions may ask which answer is **incorrect**.
- Some questions may give two or more correct answers and ask which answer is the **best**; you must consider each answer very carefully before making your choice.
- If you don't know the answer, try to work it out by **eliminating** the incorrect answers. Never leave a question unanswered.

Paper 02 (2 ½ hours)

Paper 02 is divided into **Sections A** and **B**, and consists of **six compulsory questions**, each divided into several parts. Take time to **read the entire paper** before beginning to answer any of the questions.

- **Section A** consists of **three** compulsory **structured questions** whose parts require short answers, usually a word, a sentence or a short paragraph. The answers are to be written in **spaces** provided on the paper. These spaces indicate the length of answer required and answers should be restricted to them.
    - Question 1 is a **data analysis question** which is worth **25 marks**. You will be provided with some form of **data**, such as the results obtained during a practical investigation, which you will be expected to answer questions about. The data might be in the form of a table or a graph. If you are given a table, you may be asked to draw a graph using the data and may then be asked questions about the graph. The question might also test your planning and designing skills.
    - Questions 2 and 3 are each worth **15 marks**. They usually begin with some kind of stimulus material, very often a diagram, which you will be asked questions about.
- **Section B** consists of **three** compulsory **extended response questions**, each worth **15 marks**. These questions require a greater element of **essay** writing in their answers than those in section A. One or more questions may require a **drawing** as part of your answer. It is important that you can reproduce the drawings you have been taught.

The marks allocated for the different parts of each question are clearly given. A total of **100 marks** is available for Paper 02 and the time allowed is **150 minutes**. You should allow about 35 minutes for the data analysis question worth 25 marks and allow about 20 minutes for each of the other questions. This will allow you time to read the paper fully before you begin and time to check over your answers when you have finished.

Successful revision

The following should provide a guide for **successful revision**.

- **Begin your revision early**. You should start your revision at least two months before the examination and should plan a **revision timetable** to cover this period. Plan to revise in the evenings when you don't have much homework, at weekends, during the Easter vacation and during study leave.
- When you have a **full day** available for revision, consider the day as three sessions of about three to four hours each, **morning**, **afternoon** and **evening**. Study during two of these sessions only, do something non-academic and relaxing during the third.
- **Read through the topic** you plan to learn to make sure you **understand** it before starting to learn it; understanding is a lot safer than thoughtless learning.
- Try to understand and learn **one topic** in each revision session, more if topics are short and fewer if topics are long.
- **Revise every topic** in the syllabus. Do not pick and choose topics since **all questions** on your exam paper are **compulsory**.
- **Learn the topics in order**. When you have learnt **all** topics **once**, go back to the first topic and begin again. Try to cover each topic **several times**.
- **Revise in a quiet location** without any form of distraction.
- **Sit up to revise**, preferably at a table. Do not sit in a comfy chair or lie on a bed where you can easily fall asleep.
- Obtain copies of **past CSEC® Biology examination papers** and use them to practise answering exam style questions, starting with the most recent papers. These can be purchased online from the CXC® Store.
- You can use a variety of different **methods** to **learn** your work. Chose which ones work best for you.
  - **Read the topic several times**, then close the book and try to write down the **main points**. Do not try to memorise your work word for word since work learnt by heart is not usually understood and most questions test **understanding**, not just the ability to repeat facts
  - **Summarise** the **main points** of each topic on **flash cards** and use these to help you study.
  - **Draw simple diagrams** with **annotations**, **flow charts** and **spider diagrams** to summarise topics in visual ways which are easy to learn.
  - **Practise drawing** and **labelling diagrams** that you have been given. Copy them from the book at first and then try to redraw them without the book. At least one question usually tests your ability to reproduce a drawing.
  - **Use memory aids** such as:
    - **acronyms**, e.g. **GRIMNER** for the seven life processes; **g**rowth, **r**eproduction, **i**rritability, **m**ovement, **n**utrition, **e**xcretion, **r**eproduction.
    - **mnemonics**, e.g. 'some **m**en **n**ever **p**lay **c**ricket **p**roperly' for the six major mineral elements required by plants; **s**ulfur, **m**agnesium, **n**itrogen, **p**hosphorus, **c**alcium, **p**otassium.

- **associations between words**, e.g. **tri**cuspid - **ri**ght (therefore the bicuspid valve must be on the left side of the heart), **a**rteries - **a**way (therefore veins must take blood towards the heart).

- **Test yourself** using the questions throughout this book and others from past CSEC® examination papers.

Successful examination technique

- **Read the instructions** at the start of each paper very carefully and do **precisely** what they require.
- **Read through the entire paper** before you begin to answer any of the questions.
- **Read each question at least twice** before beginning your answer to ensure you **understand** what it asks.
- **Underline the important words** in each question to help you answer precisely what the question is asking.
- **Reread** the question when you are **part way through** your answer to check that you are answering what it asks
- **Give precise** and **factual answers.** You will not get marks for information which is 'padded out' or irrelevant. The number of marks awarded for each answer indicates how long and detailed it should be.
- **Use correct scientific terminology** throughout your answers.
- Give any **numerical answer** the appropriate **unit** using the proper abbreviation/ symbol e.g. cm^3, g, °C.
- If a question asks you to give a **specific number of points**, use **bullets** to make each separate point clear.
- If you are asked to give **similarities** and **differences**, you must make it clear which points you are proposing as similarities and which points as differences. The same applies if you are asked to give **advantages** and **disadvantages**.
- **Watch the time** as you work. Know the time available for each question and stick to it.
- **Check over your answers** when you have completed all the questions.
- **Remain in the examination room** until the **end** of the examination and recheck your answers again if you have time to ensure you have done your very best. Never leave the examination room early.

Some key terms used on examination papers

Account for: provide reasons for the information given.

Annotate: add brief notes to the labels of drawings to describe the structure and/or the function of the structures labelled.

Compare: give similarities and differences.

Construct: draw a graph, histogram, bar chart, pie chart or table using data provided or obtained.

Contrast: give differences.

Deduce: use data provided or obtained to arrive at a conclusion.

Define: state concisely the meaning of a word or term.

Describe: provide a detailed account which includes all relevant information.

Discuss: provide a balanced argument which considers points both for and against.

Distinguish between or **among**: give differences.

Evaluate: determine the significance or worth of the point in question.

Explain: give a clear, detailed account which makes given information easy to understand and provides reasons for the information.

Give an account of: give a written description which includes all the relevant details.

Give an illustrated account of: give a written description which includes diagrams referred to in the description.

Illustrate: make the answer clearer by including examples or diagrams.

Justify: provide adequate grounds for your reasoning.

Outline: write an account which includes the main points only.

Predict: use information provided to arrive at a likely conclusion or suggest a possible outcome.

Relate: show connections between different sets of information or data.

State or **list**: give brief, precise facts without detail.

Suggest: put forward an idea.

Tabulate: construct a table to show information or data which has been given or obtained.

Drawing tables and graphs

Tables

Tables can be used to record numerical data and observations. When drawing a table:

- **Neatly enclose** the table and draw vertical and horizontal **lines** to separate columns and rows.
- When drawing **numerical tables**, give the correct column headings which state the **physical quantities** measured and give the correct **units** using proper abbreviations/symbols, e.g. cm^3, g, °C.
- Give the appropriate number of **decimal places** when recording numerical data.

- When drawing **non-numerical tables**, give the correct column headings and **all observations**.
- Give the table an appropriate **title** which must include reference to the responding variable and the manipulated variable.

Graphs

Graphs are used to display numerical data. When drawing a graph:

- Plot the **manipulated variable** on the **x-axis** and the **responding variable** on the **y-axis**.
- Choose appropriate **scales** which are easy to work with and which use as much of the graph paper as possible.
- Enter **numbers** along the axes and **label** each axis, including relevant units, e.g. cm^3, g, °C.
- Use a **small dot** surrounded by a small circle to plot each point.
- Plot each point **accurately**.
- Join the points with a **sharp continuous line**.
- Give the graph an appropriate **title** which must include reference to the responding variable and the manipulated variable.

Biological drawings

Drawing is one of the skills assessed for SBA. Any **biological drawing** should be:

- **Large** enough to show all structures clearly, however, space must be left at one or both sides for labels.
- Drawn using a **sharp HB pencil**, preferably a mechanical pencil with a 0.5 mm lead.
- Drawn with **single, sharp, continuous lines** which are all of **even thickness**. Lines should not be sketchy and drawings should not be shaded or coloured.
- An **accurate representation** of the specimen. It must show structures typical of the specimen but should not contain unnecessary detail. If a large number of small, repetitive structures are present, only a **few** should be drawn to show accurate detail.
- **Correctly proportioned.**
- **Labelled fully**. Label lines should be drawn using a **pencil** and **ruler**. As far as possible label lines should be **horizontal**, they should not cross and should begin **in** or **on** the structure being labelled. Labels should be neatly **printed** in **pencil** and appropriately **annotated.** If only a few structures are labelled, all labels should be on the **right**.
- **Appropriately titled**. The title must be neatly **printed** in **pencil** below the drawing and be **underlined.** The title should include the **view** or **type of section**, the **name** of the specimen or structure, and the **magnification** of the drawing.

School-Based Assessment (SBA)

School-Based Assessment (SBA) is an integral part of your CSEC® examination. It assesses you in the **Experimental Skills** and **Analysis and Interpretation** involved in laboratory and field work, and is worth **20%** of your final examination mark.

- The assessments are carried out at your school by **your teacher** during Terms 1 to 5 of your two-year programme.
- The assessments are carried out during **normal practical classes** and not under examination conditions. You have every opportunity to gain a high score in each assessment if you make a **consistent effort** throughout your two-year programme.
- Assessments are made of the following **five skills**:
  - Manipulation and Measurement
  - Observation, Recording and Reporting
  - Planning and Designing
  - Drawing
  - Analysis and Interpretation

As part of your SBA, you will also carry out an **Investigative Project** during the second year of your two-year programme. This project assesses your **Planning and Designing,** and **Analysis and Interpretation** skills. If you are studying two or three of the single science subjects, Biology, Chemistry and Physics, you may elect to carry out ONE investigation only from any one of these subjects.

You will be required to keep a practical workbook in which you record all of your practical work and this may then be moderated externally by CXC®.

1 An introduction to living organisms

Biology is the study of **living organisms.** All living organisms from the simplest unicellular organisms to the most complex multicellular organisms share certain characteristics.

The characteristics of living organisms

Living organisms have **seven** characteristics in common:

- ***Nutrition (feeding):*** *the process by which living organisms obtain or make food.*

  Animals take in ready-made food and are called **heterotrophs.** Plants make their own food and are called **autotrophs.**
- ***Respiration:*** *the process by which energy is released from food by all living cells.*

  Aerobic respiration requires oxygen and takes place in most cells. **Anaerobic respiration** takes place without oxygen in certain cells.
- ***Excretion:*** *the process by which waste and harmful substances, produced by the body's metabolism, are removed from the body.*
- ***Movement:*** *a change in the position of a whole organism or of parts of an organism.*

  Most animals can move their whole bodies from place to place. Plants and some animals can only move parts of their bodies.
- ***Irritability (sensitivity):*** *the ability of organisms to detect and respond to changes in their environment or within themselves.*
- ***Growth:*** *a permanent increase in the size and complexity of an organism.*
- ***Reproduction:*** *the process by which living organisms generate new individuals of the same kind as themselves.*

  Asexual reproduction requires only one parent. **Sexual reproduction** requires two parents.

Classification of living organisms

Using similarities and differences between living organisms they can be **classified** into groups. Simple classifications can be done based on **visible characteristics** such as number of legs, number of body parts, number of wings, presence or absence of antennae, hairiness, shape, arrangement of veins in a leaf or an insect's wing.

Scientists also use internal structures, developmental patterns, life cycles and electron microscopic techniques to classify organisms. In addition, the **modern classification** uses the molecular structure of **deoxyribonucleic acid (DNA)** to assist in grouping organisms; the greater the similarity in their DNA structure, the more closely related are the organisms.

The basic category of classification is the **species.** A species is a group of organisms of common ancestry that closely resemble each other and are normally capable of interbreeding to produce fertile offspring.

Closely related species are then grouped into **genera** (singular **genus**). Related genera are then grouped into **families**, related families into **orders**, orders into **classes**, classes into **phyla** and phyla are grouped into **kingdoms.**

i.e. species ⟶ genera ⟶ families ⟶ orders ⟶ classes ⟶ phyla ⟶ kingdoms

There are **five** kingdoms in the modern classification. Members of the kingdom **Prokaryotae** have cells that lack true membrane-bound nuclei, so their DNA is free in the cells. Members of the other four kingdoms have cells that contain true nuclei surrounded by membranes (see Chapter 5). These are known collectively as **eukaryotes.**

Viruses make up a group of organisms without any cellular structure. They are particles made up of a piece of DNA or RNA surrounded by a protein coat and they can only reproduce inside other living cells. Viruses are not included in the five kingdom classification.

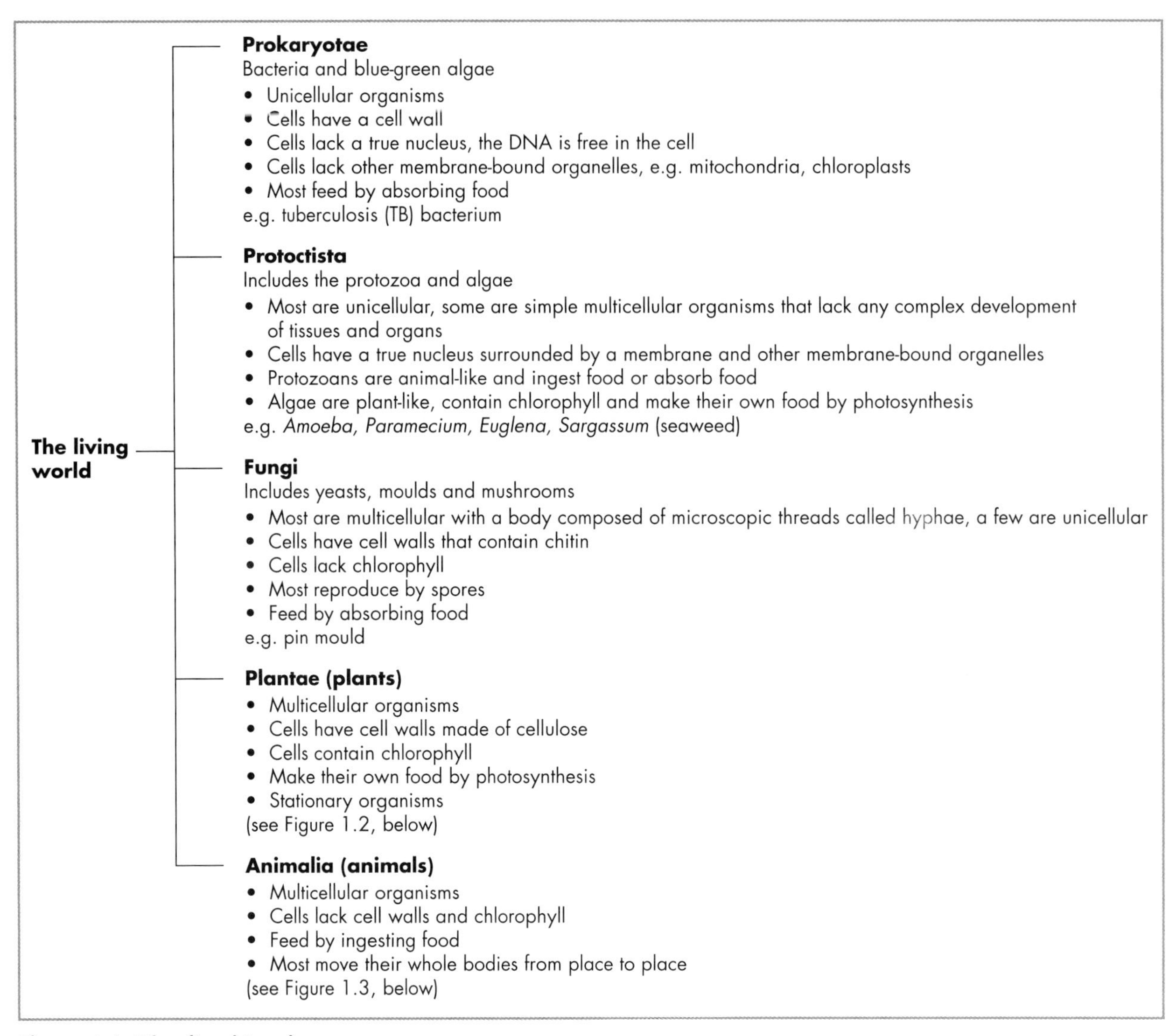

Figure 1.1 *The five kingdoms*

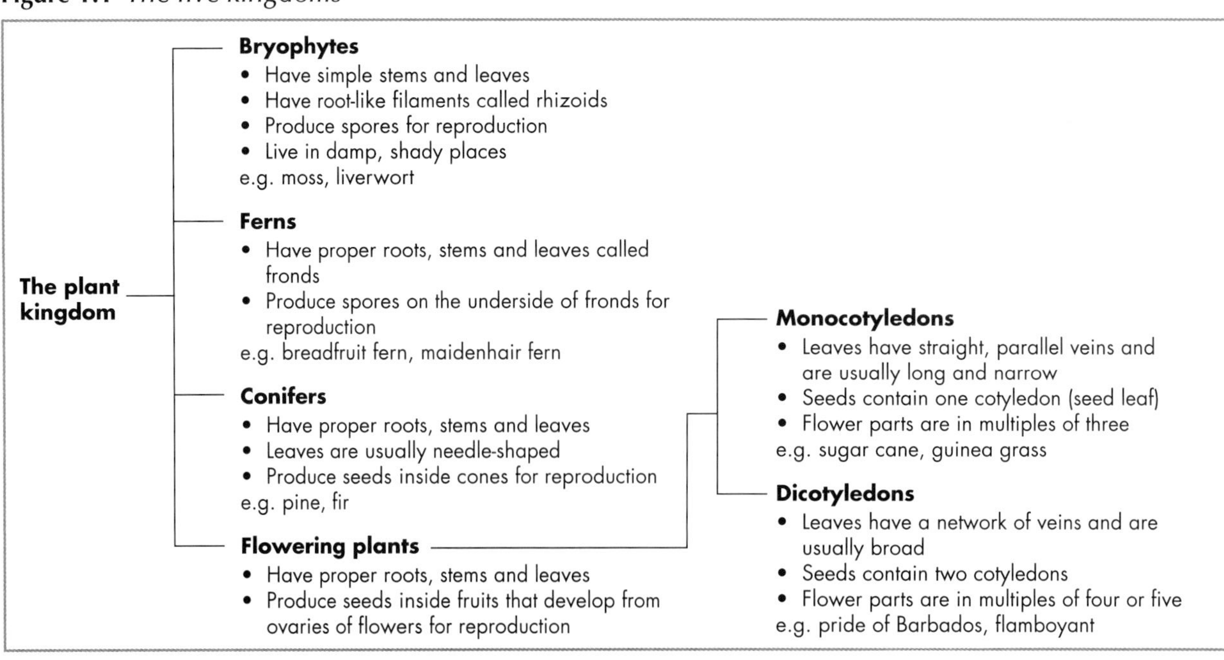

Figure 1.2 *The main groups of plants*

The animal kingdom

Porifera (sponges)
- Stationary organisms
- Lack tissues and organs
- Body contains a single cavity with many pores in its walls making a system of water canals

e.g. barrel sponge, vase sponge

Cnidaria
- Have a bag- or umbrella-shaped body
- Gut has only one opening, the mouth
- Have a ring of tentacles around the mouth

e.g. sea anemone, coral, jellyfish

Platyhelminthes (flatworms)
- Have a long, flat, unsegmented body

e.g. tapeworm, liver fluke

Nematodes (roundworms)
- Have an elongated, round, unsegmented body with pointed ends

e.g. hookworm, threadworm

Annelids (segmented worms)
- Have an elongated body divided into segments

e.g. earthworm

Arthropods
- Have a waterproof exoskeleton (external skeleton) made mainly of chitin
- Have a segmented body
- Have several pairs of jointed legs

Crustaceans
- Have two pairs of antennae
- Usually have five or seven pairs of legs
- Body is divided into head, thorax and abdomen or cephalothorax and abdomen

e.g. lobster, shrimp, crab, woodlouse

Arachnids
- Have no antennae
- Have four pairs of legs
- Body is divided into cephalothorax and abdomen

e.g. spider, tick, scorpion

Insects
- Have one pair of antennae
- Have three pairs of legs
- Body is divided into head, thorax and abdomen
- Have a pair of compound eyes
- Most have two pairs of wings

e.g. cockroach, moth, ant, house fly

Myriapods
- Have one pair of antennae
- Have many pairs of legs
- Body is elongated and divided into many segments

e.g. centipede, millipede

Molluscs
- Have a soft, moist, unsegmented body
- Have a muscular foot
- Many have shells

e.g. slug, snail, octopus

Echinoderms
- Have a body based on a radial pattern of five parts
- Body wall contains calcium carbonate plates, often with projecting spines
- Have tube feet with suction pads for movement

e.g. starfish, sand dollar, sea urchin

Chordates
- Have a notochord (rod) running down the body, most have a backbone
- Have a dorsal nerve cord with the anterior end usually enlarged forming the brain
- Most have an internal skeleton of bone and cartilage

Pisces (fish)
- Have a waterproof skin covered with scales
- Have gills for breathing
- Have fins for swimming

e.g. barracuda, flying fish, shark

Amphibians
- Have a soft, moist, non-waterproof skin without scales
- Eggs are laid in water, larvae live in water, adults live on land
- Larvae have gills, adults have lungs

e.g. frog, toad, newt

Reptiles
- Have a dry, waterproof skin with scales
- Lay eggs with a rubbery shell on land

e.g. snake, lizard, iguana, turtle

Aves (birds)
- Have a waterproof skin with feathers
- Have a beak and no teeth
- Forelimbs are modified to form wings
- Lay eggs with a hard shell
- Are homeothermic (warm blooded)

e.g. sparrow, cattle egret, hawk

Mammals
- Have a waterproof skin with hair and sweat glands
- Have different types of teeth
- Young feed on milk from their mother
- Are homeothermic

e.g. mouse, whale, human

Figure 1.3 *The main groups of animals*

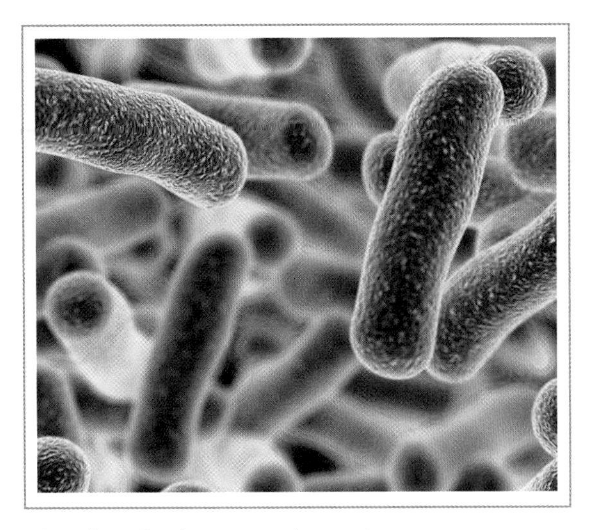

Kingdom Prokaryotae: bacteria

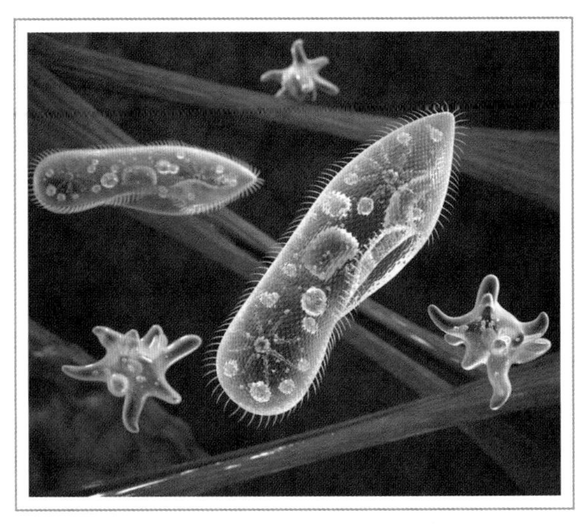

Kingdom Protoctista: Amoeba and Paramecium

Kingdom Protoctista: Sargassum (seaweed)

Kingdom Fungi: pin mould on bread

Revision questions

1. Identify FIVE ways you could use to determine if something is living.
2. What methods do scientists use to classify living organisms?
3. **a** What makes members of the kingdom Prokaryotae different from members of the other four kingdoms?

   b Name the other FOUR kingdoms into which organisms can be classified.
4. State THREE differences between plants and animals.
5. What is a species?
6. Give TWO ways to distinguish between a monocotyledon and a dicotyledon.
7. Give TWO characteristics of the members of EACH of the following groups:

   a insects **b** fish **c** mammals **d** arthropods.

2 Living organisms in their environment

Living organisms constantly interact with each other and the environment in which they live.

Definitions

- ***Ecology:*** *the study of the interrelationships of living organisms with each other and with their environment.*
- ***Environment:*** *the combination of factors that surround and act upon an organism.*

  These factors can be divided into **two** groups:
  - **Biotic factors:** all the other **living** organisms that are present such as predators, prey, competitors, parasites and pathogens.
  - **Abiotic factors:** all the **non-living** chemical and physical factors (see page 8).
- ***Habitat:*** *the place where a particular organism lives.*

  For example, the habitat of an earthworm is the upper layers of the soil.
- ***Species:*** *a group of organisms of common ancestry that closely resemble each other and are normally capable of interbreeding to produce fertile offspring.*
- ***Population:*** *all the members of a particular species living together in a particular habitat.*

  For example, all the sea urchins living in a sea grass bed form a population.
- ***Community:*** *all the populations of different species living together in a particular habitat.*

  For example, a woodland community consists of all the plants, animals and decomposers that inhabit the wood.
- ***Ecosystem:*** *a community of living organisms interacting with each other and with their abiotic environment.*

  Examples of ecosystems include a pond, a coral reef, a mangrove swamp, a grassland and a forest.
- ***Niche:*** *the position or role of an organism within an ecosystem.*

  For example, an earthworm's niche is to burrow through the soil, improving its aeration, drainage and fertility.

Carrying out an ecological study

The **aim** of studying any ecosystem is to **identify** the different species of plants and animals present, to find out **where** they live, determine their **numbers**, and find out about the **relationships** they have with each other and with the abiotic factors. Ecosystems studied could include a pond, a piece of wasteland, a small area of woodland, an area of grassland, a sand dune or a rocky shore.

Collecting organisms

To **identify** organisms, they may need to be collected. Organisms must never be collected or destroyed unnecessarily; as **few** as possible should be collected and **returned** to their original positions if possible, and their habitat should be left as **undisturbed** as possible. Pooters, pitfall traps, nets, plankton nets and a Tullgren funnel may be used.

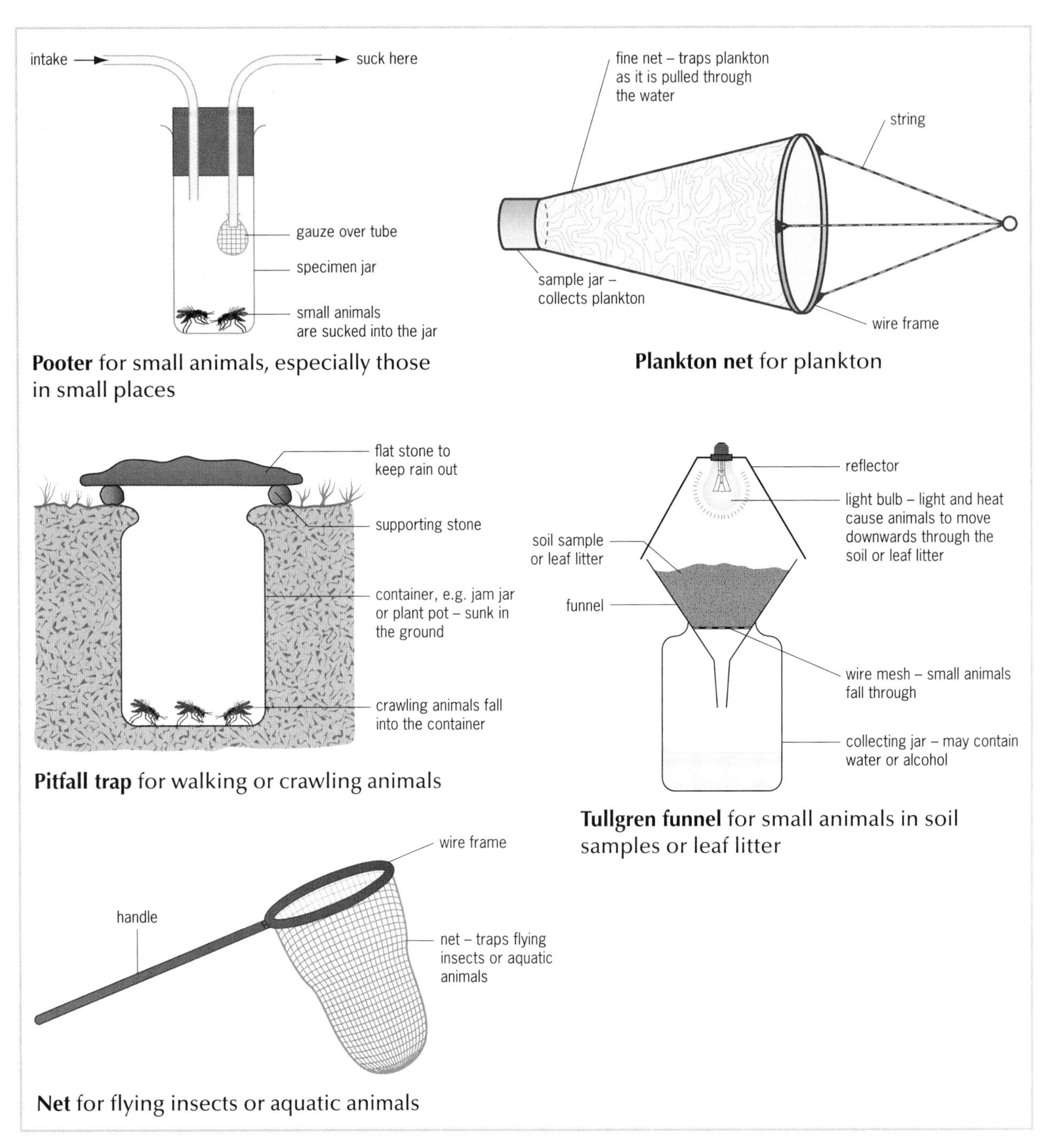

Pooter for small animals, especially those in small places

Plankton net for plankton

Pitfall trap for walking or crawling animals

Tullgren funnel for small animals in soil samples or leaf litter

Net for flying insects or aquatic animals

Figure 2.1 *Methods to collect organisms*

Sampling techniques

It is not practical to find and count all members of all species present, so **sampling techniques** are used to sample small areas from which conclusions can be drawn about the ecosystem as a whole.

- **Observation**

  The ecosystem should first be **observed** and the common species of plants and animals recorded. Any **adaptations** that enable the organisms to survive in the ecosystem should be noted, together with any **interrelationships** between the organisms.

- **Quadrats**

  A **quadrat** is a **square frame** whose area is known, e.g. 0.25 m^2 or 1 m^2. It is placed, at **random**, several times within the ecosystem. The **number** of individuals of each species of plant and stationary or slow-moving animal found within the quadrat is counted. If it is not possible to distinguish individual plants of a species, e.g. grass, the quadrat can be made into a **grid** using string and the **percentage** of the quadrat area covered can be estimated.

  Quadrats are used to study the **distribution** and **abundance** of plants and stationary or slow-moving animals in **uniform** ecosystems, e.g. an area of grassland.

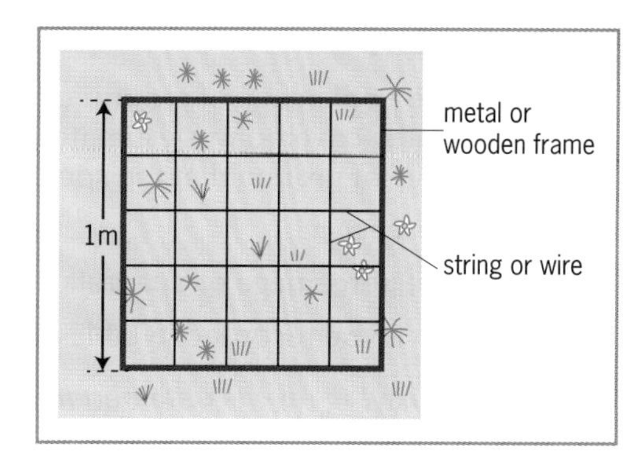

Figure 2.2 *A grid quadrat*

- **Line transects**

  A **line transect** is usually a measuring tape or string that has marks at regular intervals, e.g. 10 cm or 25 cm. It is placed in a **straight line** across the ecosystem and the species of plants and stationary or slow-moving animals touching the line, or touching the line at each mark, are recorded.

  Line transects are useful where there is a **transition** of organisms across the ecosystem, e.g. down a rocky seashore. They give a quick idea of the **species present** and how they **change** across the ecosystem.

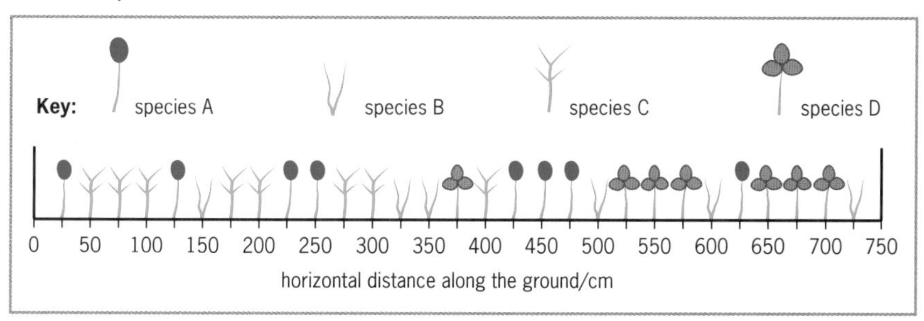

Figure 2.3 *Displaying the results of a line transect for four species, A, B, C and D*

- **Belt transects**

  A **belt transect** is a strip of fixed width, e.g. 0.5 m or 1 m, made by placing two parallel line transects across an ecosystem. The species found between the lines are recorded. Alternatively, a **quadrat** can be placed alongside one line transect and the **number** of individuals of each species found within its boundaries is counted. The quadrat is then moved along the line and counting is repeated at regular intervals.

Estimating population sizes

Population sizes can be estimated by using the results from quadrats or by using the capture–recapture method.

- **Using results from quadrats**

  Results from quadrats can be used to obtain:

  - **Species density**

    This is the average number of individuals of a given species per m^2. If the quadrat is 1 m^2, it is obtained by dividing the total number of individuals of the species by the number of quadrats used.

  - **Total population**

    This is the total number of individuals of a given species in the area under study. It is obtained by multiplying the species density by the total area of the ecosystem studied.

- **Species cover**

  This is the percentage of ground covered by a given species. It is used if the percentage of the quadrat area covered was estimated. If the quadrat is 1 m^2, it is obtained by dividing the total percentage of ground that the species covered by the number of quadrats used.

- **Species frequency**

  This is the percentage of quadrats in which the given species was found.

These results can be recorded in **tables** and **bar charts**.

Table 2.1 *Displaying results obtained for four species using a 1 m^2 quadrat*

Species	Number of organisms or percentage cover in each quadrat (Q)										Total of 10 quadrats	Species density/ number of individuals per m^2	Species cover/%	Species frequency/%
	Q1	Q2	Q3	Q4	Q5	Q6	Q7	Q8	Q9	Q10				
W	6	8				4			7	5	30	3		50
X	12		5	15	7	12	8	4	16	1	80	8		90
Y	10%	50%	45%	5%	40%	30%		5%		25%	210%		21	80
Z	15%			25%	5%		30%	10%	5%		90%		9	60

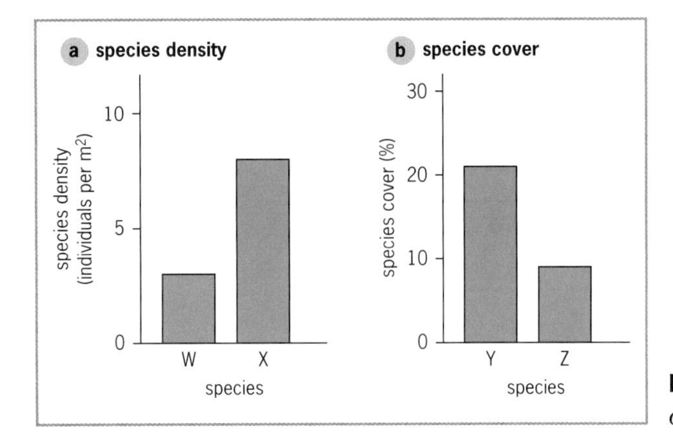

Figure 2.4 *Bar charts displaying the results obtained from quadrats*

- **Capture–recapture method**

  Capture–recapture is used to estimate **population sizes** of moving animals. A representative sample of animals of a species is collected, the animals are counted and each one is **marked**, e.g. with a dot of paint. The animals are released back into the same area and given time to mix with the original population. A second sample is collected, counted and the number of marked organisms counted. The population size is then estimated:

$$\text{estimated population size} = \frac{\text{number of organisms in first sample} \times \text{number of organisms in second sample}}{\text{number of marked organisms recaptured}}$$

The impact of abiotic factors on living organisms

The abiotic environment supplies living organisms with many of the **requirements** that they need to survive. **Abiotic factors** are important in determining the **types** and **numbers** of organisms that exist in a given environment, i.e. they influence the **distribution** and **behaviour** of living organisms. These factors can be divided into the following:

- **edaphic** factors
- **climatic** factors
- **aquatic** factors.

Edaphic factors

Edaphic factors are those connected with the **soil**. Soil is a mixture of inorganic rock particles, water, air, mineral salts, organic matter and living organisms including plant roots, small animals and micro-organisms. Soil provides organisms living in it with water, oxygen, nitrogen and mineral nutrients.

Table 2.2 *Edaphic factors and their impact on living organisms*

Edaphic factor	Impact on living organisms
Texture, i.e. composition of rock particles Rock particles form from rocks by weathering and are classified by size: **Clay:** less than 0.002 mm **Silt:** 0.002–0.02 mm **Sand:** 0.02–2.0 mm **Gravel:** greater than 2.0 mm The percentage of each determines the texture of the soil.	• Influences the **water** content of the soil. The smaller the soil particles, the more water held by capillarity and chemical forces and the higher the water content. • Influences the **air** content of the soil. The larger the soil particles, the larger the air spaces and the higher the air content. • Influences the **mineral ion** content of the soil. The smaller the soil particles, the harder it is to leach and the higher the mineral content. • Influences how easy it is for **plant roots** to penetrate and **animals** to burrow through the soil. The larger the particles, the easier to penetrate and burrow through.
Water Obtained from rainfall and held in a thin film around soil particles by capillarity and chemical forces.	• Essential for **photosynthesis** in plants. • Dissolves **minerals** so they can be absorbed by plant roots. • Prevents the **desiccation** (drying out) of soil organisms without waterproof body coverings, e.g. earthworms.
Air Present in the spaces between the soil particles.	• **Oxygen** in the air is essential for **aerobic respiration** in plant roots and soil organisms. • **Oxygen** in the air is necessary for bacteria and fungi to **decompose** organic matter aerobically to form **humus**. • **Nitrogen** in the air is necessary for **nitrogen fixing bacteria** to form inorganic **nitrogenous compounds**, e.g. nitrates.
Mineral nutrients Formed from decomposing organic matter and by dissolving from the surrounding rock. Present as **ions** dissolved in soil water.	• Essential for **healthy growth** of plants (see Table 7.1, page 50)
pH The optimum pH for most plants is 6.0 to 7.5	• Mainly affects **mineral ion** availability for plants. If the soil is too acidic or alkaline, mineral ions become less available.
Humus Formed by bacteria and fungi decomposing dead or waste **organic matter**. Humus is a dark brown, sticky material which coats soil particles, mainly in the **topsoil**.	• Improves the **air content** by binding soil particles together in small clumps called **soil crumbs**. • Improves the **mineral ion** content by adding minerals, and absorbing and retaining minerals. • Improves the **water content** by absorbing and retaining water.

Climatic factors

The **climatic factors** affecting **terrestrial** organisms include light, temperature, humidity, water availability, wind and atmospheric gases.

- **Light**

  Light intensity and its **duration** affect living organisms. Light is essential for plants to make food by **photosynthesis** and the rate of photosynthesis depends on light intensity. Light also synchronises activities of plants and animals with the **seasons**, e.g. flowering in plants, and migration, hibernation and reproduction in animals.

- **Temperature**

  Most organisms can only survive within a certain, **narrow temperature range**. At low temperatures, ice crystals may form in cells and damage them. At high temperatures, enzymes are denatured. Temperature affects the rate of **photosynthesis** and **germination** in plants and the **activity** of animals, e.g. many animals become dormant in low temperatures.

- **Humidity**

  Humidity affects the rates of **transpiration** in plants and **evaporation** of water from some animals.

- **Water availability**

  Terrestrial habitats receive water by precipitation from the atmosphere as rain, snow, sleet and hail. Water is essential for life. **Chemical reactions** in cells and most **life processes** need water in order to take place (see page 54). It is also essential for **photosynthesis** in plants.

- **Wind**

  Wind is essential for **pollination** and **seed dispersal** in many plants and may also influence **migration** of birds. It can affect the rate of **transpiration** in plants, the rate of **evaporation** of water from animals and the **growth** of vegetation, e.g. branches on the windward side of trees in exposed places become stunted and deformed.

Seed dispersing in the wind

Tree exposed to the wind

- **Atmospheric gases**

  The air is a mixture of gases including nitrogen, oxygen, carbon dioxide, water vapour and pollutants. **Oxygen** is essential for **aerobic respiration** in almost all living organisms, **carbon dioxide** is essential for **photosynthesis** in plants and **pollutant gases** have a negative effect on living organisms.

Aquatic factors

Aquatic factors affect organisms living in **aquatic habitats**, e.g. ponds, lakes, rivers, oceans and coral reefs. Aquatic organisms are affected by light, temperature, water availability and pollutants in the same way as terrestrial organisms. They are also affected by other factors.

- **Salinity**

  Salinity refers to the concentration of **salt** in the water. Most aquatic organisms are adapted to survive in a **specific salinity** and can only tolerate small changes. Some are adapted to live in fresh water, some in salt (sea) water and others in brackish water, e.g. in estuaries.

- **Water movement and wave action**

  Organisms living in rivers and streams are affected by the constant movement of water, those living along ocean shorelines are exposed to wave action and to changing water levels as the tide changes, and those living in oceans are affected by currents.

- **Dissolved oxygen**

  Aquatic organisms use oxygen dissolved in the water for **respiration**. Oxygen levels in still or polluted water are often low, whereas habitats exposed to moving water usually have sufficient oxygen.

Revision questions

1. What is ecology?

2. Distinguish between the following pairs of terms:

   a habitat and niche

   b population and community

   c species and population.

3. Identify FOUR methods you could use to collect organisms from a habitat.

4. Explain how you would estimate:

   a the total population of a small, named plant growing in an area of wasteland

   b the total number of snails in a garden.

5. Define the term 'environment'.

6. Distinguish between the biotic environment and the abiotic environment.

7. Discuss the importance of EACH of the following edaphic (soil) factors to organisms living in the soil:

   a water **b** air **c** mineral nutrients.

8. Discuss the importance of EACH of the following climatic factors to living organisms:

   a light **b** temperature **c** atmospheric gases.

3 Interrelationships between living organisms

Many kinds of **relationships** exist between living organisms. These relationships may be beneficial or harmful, close or loose.

Feeding relationships

Food chains

Organisms within any ecosystem are linked to form **food chains** based on how they obtain **organic food**.

Energy from the sun enters living organisms through **photosynthesis** occurring in green plants, also known as **primary producers**. This energy is incorporated into **organic food molecules** produced by the plants and is passed on to **consumers** through food chains. A food chain includes:

- A **primary producer**, i.e. a green plant.
- A **primary consumer** which eats the primary producer.
- A **secondary consumer** which eats the primary consumer.
- A **tertiary consumer** which eats the secondary consumer.
- Some food chains may also include a **quaternary consumer** which eats the tertiary consumer.

Consumers can also be classified according to what they consume:

- **Herbivores** consume plants, e.g. cows, grasshoppers, snails, slugs, parrotfish, sea urchins.
- **Carnivores** consume animals, e.g. lizards, toads, spiders, centipedes, eagles, octopuses, sharks.
- **Omnivores** consume both plants and animals, e.g. hummingbirds, crickets, humans, crayfish.

The levels of feeding within a food chain are referred to as **trophic levels**.

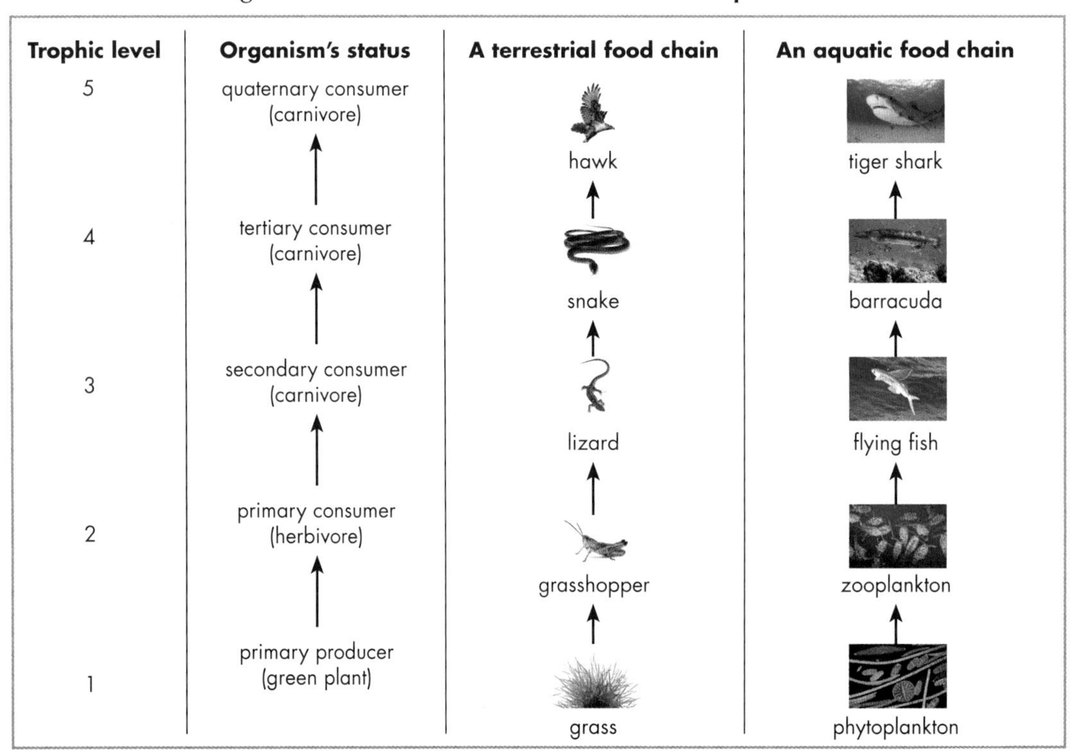

Figure 3.1 *Examples of food chains*

Predator/prey relationships

A **predator** is an organism that kills and feeds on another organism, e.g. a lion. The **prey** is the organism that the predator eats, e.g. a zebra. Predators are often the prey of other organisms and prey are often predators themselves. An **apex predator** is at the top of a food chain and has no predators, e.g. killer whales, tiger sharks, hawks, owls.

Table 3.1 *Predator/prey relationships from Figure 3.1*

Habitat	Predator	Prey
Terrestrial	hawk snake lizard	snake lizard grasshopper
Aquatic	tiger shark barracuda flying fish	barracuda flying fish zooplankton

A hawk swooping towards its prey

A predator and its prey **evolve** together. The **predator** must evolve characteristics to **catch its prey**, e.g. speed, stealth, camouflage, highly developed senses, sharp and piercing mouthparts, poison to kill its prey, immunity to the prey's poison. At the same time, the **prey** evolves characteristics to **avoid being eaten**, e.g. speed, camouflage, highly developed senses, rapid responses, poison, protective body coverings.

Predator/prey relationships serve as biological controls and keep the numbers of organisms in an ecosystem **relatively constant**. If a predator overhunts its prey, the prey population will decrease and this will cause the predator population to decrease. The prey population will then begin to increase again, which will allow the predator population to increase.

Humans can use predator/prey relationships to **control pests**. This is known as **biological control**. It involves introducing a natural predator of the pest into the environment, e.g. mongooses were introduced into Barbados to control the snake population.

Food webs

Any ecosystem usually has more than one primary producer and most consumers have more than one source of food. Consequently, food chains are interrelated to form **food webs**.

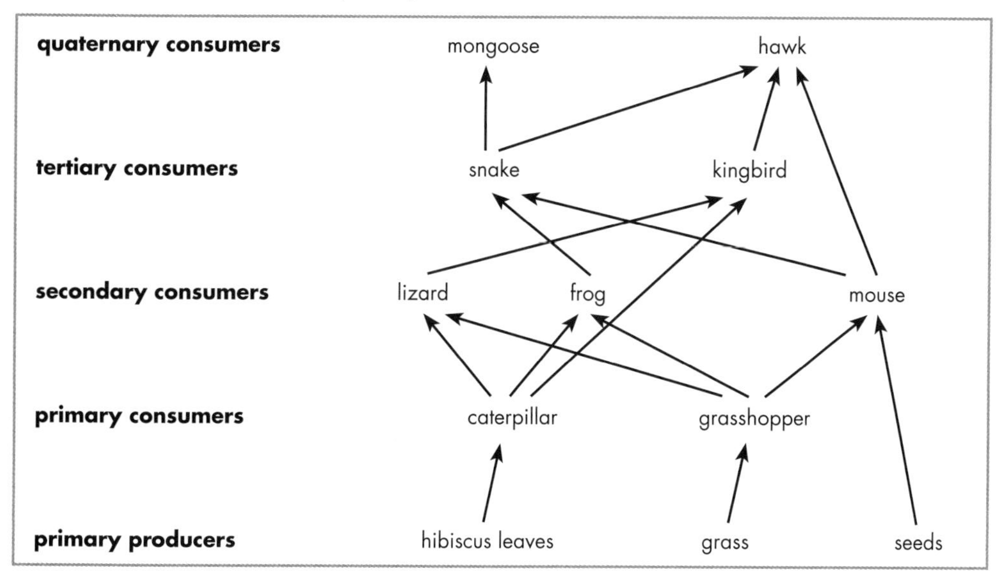

Figure 3.2 *An example of a terrestrial food web*

Detritivores and decomposers

Detritivores and decomposers are organisms that are present in ecosystems. They are essential for the **recycling** of chemical elements within all ecosystems (see page 17).

- **Detritivores** are animals, e.g. earthworms, woodlice, millipedes and sea cucumbers, which feed on pieces of decomposing organic matter, breaking them down into smaller fragments.
- **Decomposers** are micro-organisms, i.e. bacteria and fungi, which feed **saprophytically** on dead and waste organic matter causing it to **decompose**. They secrete digestive enzymes that breakdown complex organic compounds into simple organic compounds which they absorb. During this process, they release **carbon dioxide** and inorganic **mineral nutrients** in the form of **ions**, e.g. nitrates and sulfates, into the environment. These can then be reabsorbed and **re-used** by plants.

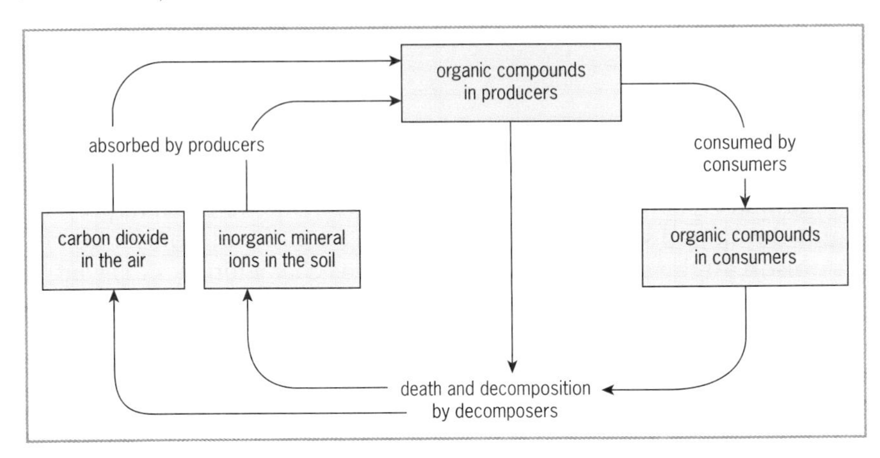

Figure 3.3 *Cycling in a terrestrial ecosystem*

Symbiotic relationships

Symbiosis is any **close relationship** between two organisms of different species. Symbiotic relationships can be divided into the following types:

- **parasitism**
- **commensalism**
- **mutualism**.

Parasitism

In **parasitism**, one organism, the **parasite**, gains **benefit** and the other organism, the **host**, is **harmed**. The parasite lives in or on the host and often has a **vector** or **intermediate host** which is unharmed by the parasite. The parasite usually reproduces inside the intermediate host and this increases the parasite's chances of survival and transmission.

- **Lice** and **ticks** live on certain mammals, e.g. humans, dogs and cows, sucking their **blood**. Ticks on cattle cause damage to the hide, weakness, anaemia and tick paralysis.
- **Tapeworms** live in the intestines of humans. The tapeworm absorbs **digested food** from the intestines and also gains shelter and protection. The infected person may suffer from abdominal pains, loss of appetite, weight loss and nausea. A pig is usually the intermediate host.
- ***Plasmodium*** is the parasite that causes **malaria** in humans. After initially reproducing in the liver, the parasites enter red blood cells where they live and reproduce.

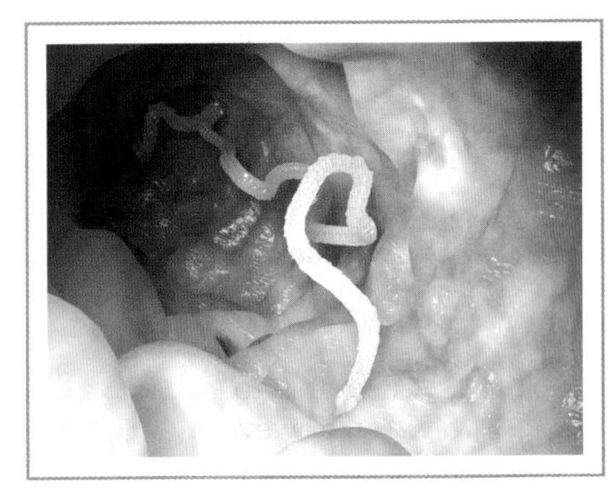

A tapeworm in human intestines

A **high fever** develops when the cells burst and release the parasites and subsides as the parasites enter more red blood cells. It then recurs as the cells burst and the cycle continues resulting in a **recurrent fever**. The *Anopheles* mosquito is the intermediate host.

- **Dodder (love vine)** is a parasitic plant with long yellow stems that twist around other plants from small shrubs to tall trees. Outgrowths from the stems penetrate into the host's phloem and absorb **sugars** and **amino acids**. This causes reduced growth of the host.

Bromeliads growing on tree trunks

Commensalism

In **commensalism**, one organism, the **commensal**, gains **benefit** while the other organism neither gains nor is harmed.

- **Epiphytes** are plants, e.g. orchids and bromeliads, which grow non-parasitically on other plants, usually **trees**, which they use for **support** since their roots do not enter the soil. They usually grow high up on the branches where they are close to sunlight, have very little shade and are out of the reach of herbivores on the ground.
- **Cattle egrets** are commensals that perch on the back of **cows**. They gain **food** by eating ticks from the cow's skin and insects that the cow disturbs as it moves through the grass.
- **Remoras** are small fish that attach themselves onto the skin of **sharks** by suction cups on their heads. They gain **food scraps** left by the sharks as they feed.

Remora attached to a shark

Mutualism

In **mutualism**, **both** organisms gain **benefit**, and in many cases, they cannot survive without each other.

- **Leguminous plants**, e.g. peas and beans, have **nitrogen fixing bacteria** living in swellings on their roots called **root nodules**. The bacteria use nitrogen in the air in the soil to produce inorganic nitrogenous compounds. The plants gain **nitrogenous compounds** which they use to manufacture proteins. The bacteria gain **food**, which the plants produce during photosynthesis, and **protection**.
- **Coral polyps** have **green algae** living within the tissues lining their digestive cavities. The polyps gain **food** and **oxygen** as the algae photosynthesise. The algae gain **carbon dioxide** as the polyps respire, **nitrogenous compounds** excreted by the polyps, and **protection**.
- **Termites** have **protozoans** living in their intestines. Termites are unable to digest the cellulose in the wood they eat; however, the protozoans can digest this cellulose into sugars. The termites gain **digested food**. The protozoans are supplied with **food** and **protection**.

Root nodules of a leguminous plant

Coral polyps

Other interrelationships

A variety of other interrelationships exist between living organisms:

- **Camouflage:** some organisms resemble others so they are concealed, e.g. stick insects resemble woody stems.
- **Pollination:** many plants depend on insects, small birds or bats to transfer their pollen from one flower to another for reproduction.
- **Support:** some organisms use others for support, e.g. birds build nests in trees, vines use the support of other plants to grow closer to sunlight.
- **Protection:** some organisms use others for protection, e.g. grasshoppers live in long grass.
- **Competition:** members of the same species and of different species may compete with each other. Animals compete for food, space, a mate and shelter. Plants compete for light, water and minerals.

Energy flow in ecosystems

During **photosynthesis**, primary producers absorb **sunlight energy** and convert it into **chemical energy**, which is stored in organic food molecules. Some of this energy is then released by the producers during **respiration** and some is passed on through **food chains** in the organic molecules.

At each trophic level in a food chain, **energy** and **biomass** (amount of biological matter) are **lost**. Some organic matter containing energy is lost in **faeces** and some is lost in organic **excretory products**, e.g. urea. Some is used in **respiration** during which the stored energy is released and used, or lost as heat. The remaining energy containing organic matter is used in growth and repair, and is then passed on to the next trophic level when organisms are consumed. Organisms not consumed eventually **die**.

The organic matter in faeces, excretory products and dead organisms is **decomposed** by decomposers and they release the energy during respiration. **Energy**, therefore, flows from producers to consumers and decomposers in **one direction** through ecosystems, and is not recycled. In general, only **10%** of the energy from one level is transferred to the next level.

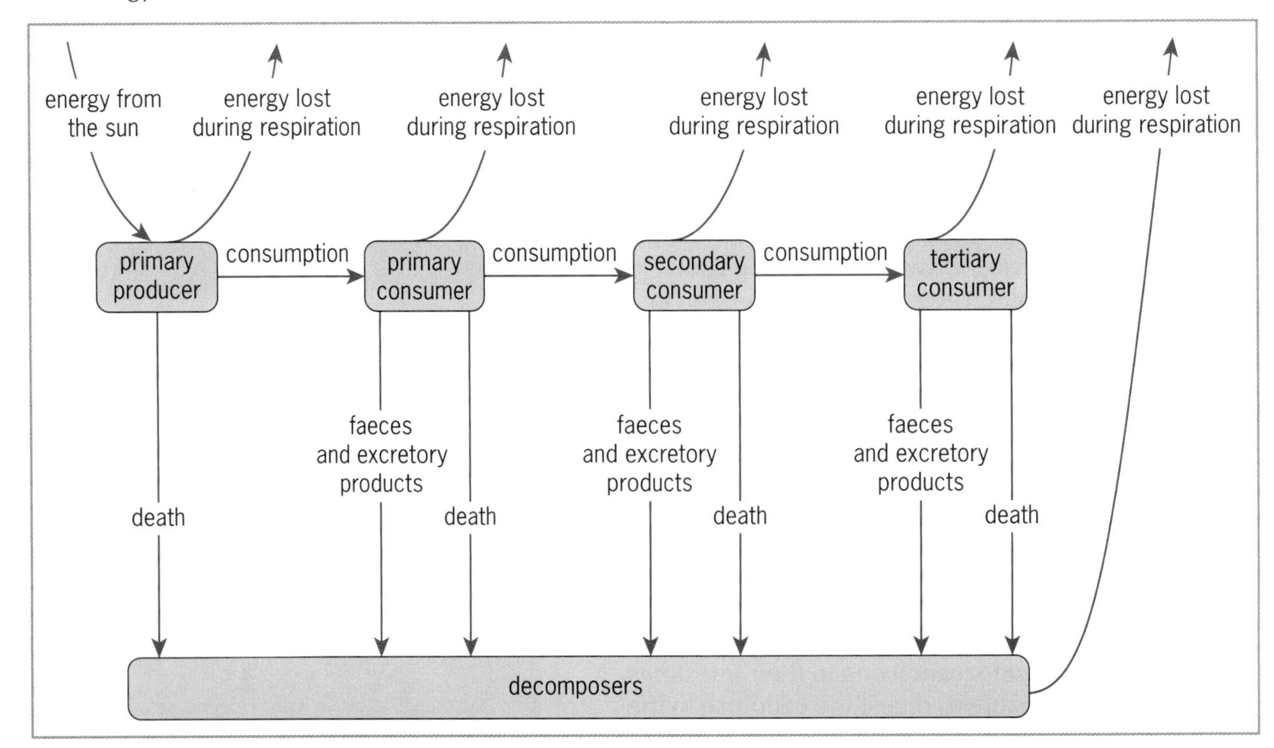

Figure 3.4 *Energy flow through an ecosystem*

Ecological pyramids

Because there is less energy and biomass at each trophic level in a food chain, fewer organisms can be supported at each level. **Energy**, **biomass** and **number of organisms** at successive levels can be represented by **ecological pyramids**. Due to the loss of energy and biomass at each level, food chains rarely exceed four or five trophic levels.

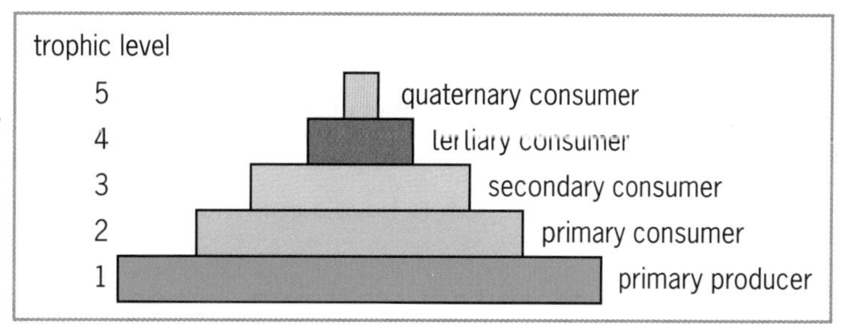

Figure 3.5 *Pyramid of energy, biomass or numbers*

Recycling

Recycling of materials in nature

Materials are constantly being **recycled** and **re-used** in nature. The different **chemical elements** that make up the bodies of all living organisms, mainly carbon, hydrogen, oxygen, nitrogen, phosphorus, sulfur and calcium, are continually **cycled** through these living organisms and the physical environment. **Decomposers** are essential to the recycling of most of these elements.

- **The cycling of water**

  The cycling of **water** is essential to ensure that:

  - Plants have a continuous supply of water to manufacture organic food by **photosynthesis**.
  - All living organisms have a continuous supply of water to keep their cells **hydrated** and to act as a **solvent** (see page 54).
  - Aquatic organisms have a constant **environment** in which to live.

- **The cycling of carbon**

  Carbon atoms are cycled by being converted into different compounds containing carbon, e.g. carbon dioxide and all organic compounds. The cycling of carbon is essential to ensure that:

  - Plants have a continuous supply of carbon dioxide to manufacture organic food by **photosynthesis**.
  - Animals and decomposers have a continuous supply of **organic food**.

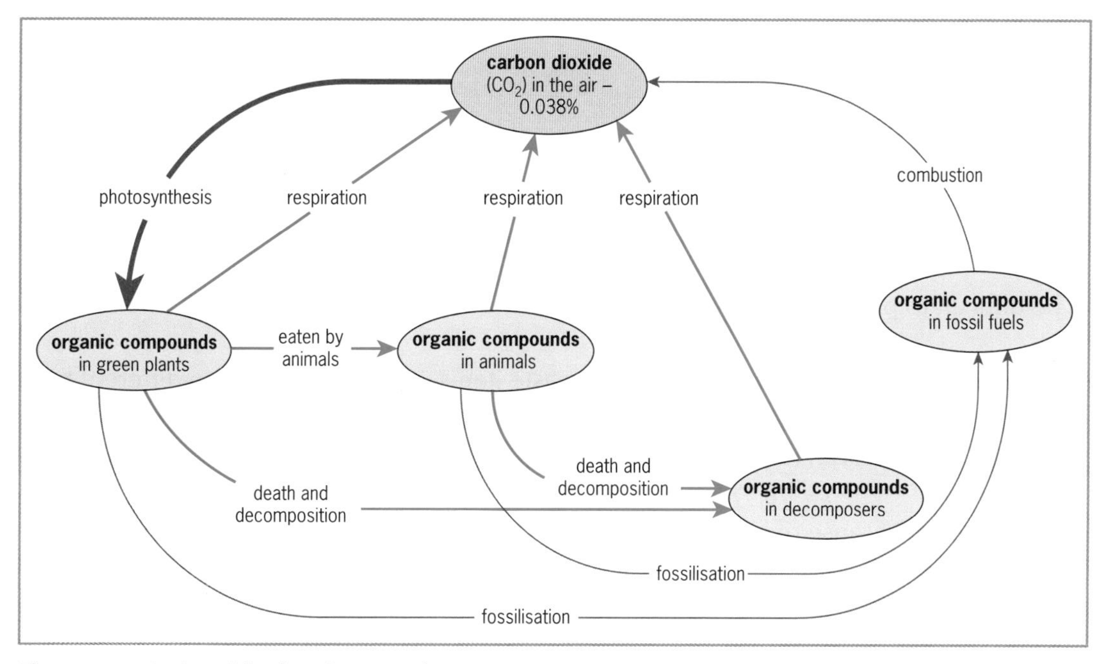

Figure 3.6 *A simplified carbon cycle*

- **The cycling of nitrogen**

  Nitrogen atoms are cycled by being converted into different compounds containing nitrogen, e.g. nitrates and proteins. The cycling of nitrogen is essential to ensure that:
  - Plants have a continuous supply of **nitrates** to manufacture proteins.
  - Animals and decomposers have a continuous supply of **proteins**.

Recycling of manufactured and other waste materials

Recycling changes waste materials into new products. **Recyclable materials** can be divided into **two** categories:

- **Biodegradable materials:** these can be decomposed by the action of living organisms, mainly bacteria and fungi. They include waste from the food industry, farmyard and garden waste, most paper and bagasse from the sugar industry.
- **Non-biodegradable materials:** these cannot be decomposed by living organisms. They include glass, plastics, rubber, construction waste, synthetic fabrics such as nylon, and metals such as iron, steel, aluminium, copper and lead.

Bagasse is biodegradable

Plastics are non-biodegradable

Recycling of manufactured and other waste materials is **important** because it:

- **Prevents wastage** of potentially useful materials.
- **Conserves natural resources** by reducing the quantity of fresh raw materials used in manufacturing.
- Reduces **energy** usage.
- Reduces the **quantity** of waste requiring disposal.
- Reduces **pollution** of air, land and water.

Difficulties encountered in recycling manufactured materials

Several **difficulties** are encountered when trying to recycle materials:

- It can be difficult to persuade households and industries to **separate** their waste into different types.
- It is more difficult to **collect**, **transport** and **store** waste items when separated into different types.
- It can be **time consuming** because items have to be cleaned before they are recycled. Also, different manufactured materials can have very different properties and they have to be sorted into their different types before recycling, e.g. there are many different types of plastics.
- It can be **hazardous** because recyclable materials have to be separated from any toxic materials before they can be recycled, e.g. the acid has to be removed from lead batteries before recycling the lead.

- It can be **uneconomical** in small countries such as the Caribbean islands because it is labour and energy intensive, and the quantity of recyclable materials generated by these countries is insufficient to maintain the full-time operation of recycling plants.
- Most small countries do not have the **facilities** to use recycled raw materials.

Revision questions

1. Some aphids were observed on the tomato plants in a garden and ladybird beetles were seen feeding on the aphids. The ladybirds were, in turn, being eaten by dragonflies which were, themselves, being fed on by toads. Use this information to draw a food chain for the organisms in the garden.

2. From the organisms in question **1**, identify:

   a a carnivore **b** a herbivore **c** a primary producer
   d a primary consumer **e** a secondary consumer **f** a predator/prey relationship.

3. What are decomposers and why are they essential within any ecosystem?

4. What is a symbiotic relationship?

5. Give ONE example of EACH of the following symbiotic relationships:

   a mutualism **b** commensalism **c** parasitism.

   For EACH example, name the partners involved and discuss the impact of the relationship on BOTH partners.

6. Explain why food chains rarely contain more than four or five trophic levels.

7. With reference to water, carbon and nitrogen, explain why it is important to continually recycle materials in nature.

8. Suggest FOUR reasons why it is important to recycle manufactured and other waste materials in today's world.

9. Suggest THREE difficulties encountered when trying to recycle manufactured and other waste materials in small countries such as the islands of the Caribbean.

4 The impact of humans on the environment

The **human population** is currently growing at about 1.2% per year. This growth, together with improved standards of living, is having a profound effect on all other living organisms, natural resources and the environment in general.

The impact of human activities on natural resources

Human activities are having a **negative impact** on both non-renewable and renewable natural resources, and in many cases, these resources are being **rapidly depleted**.

Non-renewable resources

Non-renewable resources are present in the Earth in **finite** amounts; they cannot be replaced, and consequently they are **running out**. These include:

- **Energy resources** such as fossil fuels, i.e. petroleum (crude oil), natural gas and coal, and radioactive fuels, e.g. uranium.
- **Mineral resources** such as iron ore, bauxite (aluminium ore), copper and tin.

Renewable resources

Renewable resources can be **replaced** by natural processes. However, many plant and animal species are being **overexploited** such that their numbers are decreasing, in some cases to the point of **extinction**.

- Many **marine** organisms are being **overfished** for food, e.g. lobsters, whales, turtles, sea eggs and conch.
- Some terrestrial organisms are being overhunted for products such as fur and ivory, e.g. mink, seal and elephants.

Illegal hunting endangers elephants

- Vast areas of **forest** are being **cut down** to provide land for housing and agriculture, and to provide materials for fuel, building and the manufacture of paper. This **deforestation** leads to:
  - The loss of a **habitat** for plants and animals.
  - The **destruction** of plants and animals living in the forests. Some of these may eventually become **extinct**.
  - A reduction in **photosynthesis** resulting in a gradual increase in atmospheric carbon dioxide levels which is contributing to the greenhouse effect (see page 26).
  - Disruption of the **water cycle**.
  - **Soil erosion** caused by the absence of leaves to break the force of the rain and roots to bind the soil.

Deforestation leaves soil exposed to soil erosion

- **Soil** is being **eroded** due to cutting down trees and not replanting, and bad agricultural practices such as leaving the soil barren after harvesting, using chemical fertilisers instead of organic fertilisers, overgrazing of animals and ploughing down hillsides instead of contour ploughing. The loss of soil leads to:
  - A reduction in the number of **trees** and **other plants** that can be grown.
  - A reduction in the quantity of **agricultural crops** that can be grown.

The negative impact of human activity on the environment

Many **human activities** such as agriculture, industry, mining and disposal of waste have a negative impact on living organisms and the environment. These activities:

- Cause the **destruction** and consequent **loss** of habitats and organisms living in them.
- Release **waste** and **harmful substances** into the environment which damage the environment, harm living organisms and have a negative effect on human health, i.e. they cause **pollution**.

Pollution caused by agricultural practices and industry

Modern **agricultural practices** and **industry** produce waste products that pollute the air, land and water.

***Pollution** is the contamination of the natural environment by the release of unpleasant and harmful substances into the environment.*

Table 4.1 *Pollution caused mainly by agricultural practices*

Pollutant	Origin	Harmful effects
Pesticides, e.g. insecticides, fungicides and herbicides	• Used in agriculture to control pests, diseases and weeds. • Used to control vectors of disease, e.g. mosquitoes.	• Become **higher in concentration** up food chains and can harm top consumers. • Can **harm useful** organisms as well as harmful ones, e.g. bees, which are crucial for pollination in plants.
Nitrate ions (NO_3^-) and phosphate ions (PO_4^{3-})	• Chemical fertilisers used in agriculture. • Synthetic detergents.	• Cause **eutrophication**, i.e. the rapid growth of green plants and algae in lakes, ponds and rivers, which causes the water to turn green. The plants and algae begin to die and are decomposed by aerobic bacteria that multiply and use up the dissolved oxygen. This causes other aquatic organisms to die, e.g. fish.  *Eutrophication*

Table 4.2 *Pollution caused mainly by industry*

Pollutant	Origin	Harmful effects
Carbon dioxide (CO_2)	• Burning fossil fuels in industry, motor vehicles, power stations and aeroplanes.	• Builds up in the upper atmosphere enhancing the **greenhouse effect**, which is leading to **global warming** (see page 26). • Some is also absorbed by oceans causing **ocean acidification** (see page 26).
Carbon monoxide (CO)	• Burning fossil fuels in industry and motor vehicles. • Bush fires and cigarette smoke.	• Combines with haemoglobin more easily than oxygen. This reduces the amount of oxygen reaching body cells which reduces respiration and mental awareness. It causes dizziness, headaches and visual impairment, and can lead to unconsciousness and death.
Sulfur dioxide (SO_2)	• Burning fossil fuels in industry and power stations. *Pollution from industry*	• Causes respiratory problems, e.g. bronchitis, and reduces the growth of plants. • Dissolves in rainwater forming **acid rain**. Acid rain decreases the pH of the soil, damages plants, harms animals, corrodes buildings, and causes lakes, streams and rivers to become acidic and unsuitable for aquatic organisms. • Combines with water vapour and smoke forming **smog**, which causes respiratory problems, e.g. bronchitis, asthma and lung disease.
Oxides of nitrogen (NO and NO_2)	• Combustion at high temperatures in industry, motor vehicles and power stations.	• Very toxic. Cause lung damage and even at low concentrations they irritate the respiratory system, skin and eyes. • Reduce plant growth, cause leaves to die and dissolve in rainwater forming **acid rain** (see above).
Carbon particles (smoke)	• Burning fossil fuels in industry. • Bush fires and cigarette smoke.	• Coat leaves which reduces photosynthesis, and blacken buildings. • Combine with water vapour and sulfur dioxide to form **smog** (see above).
Dust and other particulate matter	• Industry. • Mining and quarrying.	• Cause respiratory problems, e.g. bronchitis, asthma and lung disease. • Coat leaves which reduces photosynthesis.
Heavy metal ions, e.g. mercury, cadmium, lead and arsenic ions	• Burning fossil fuels in industry. • Extraction and purification of metals.	• May be directly toxic to organisms or become **higher in concentration** up food chains, harming top consumers. • Damage many body tissues and organs, especially the nervous system.

Pollution caused by the improper disposal of garbage

A very small amount of human garbage is recycled; most is dumped in landfills, garbage dumps, gullies, waterways, oceans, by the roadside, or is incinerated. **Improper disposal of garbage** is a threat to the environment:

- **Toxic chemicals** in the garbage can leach out and contaminate the soil, aquatic environments and water sources.
- **Greenhouse gases**, e.g. methane and carbon dioxide, can be released into the atmosphere where they contribute to the **greenhouse effect** (see page 26).
- **Hydrogen sulfide gas** can be released into the air. This gas is extremely toxic, and even low concentrations irritate the eyes and respiratory system.
- **Plastics** can enter waterways and oceans where they are harmful to aquatic organisms.
- **Bacteria** from untreated sewage can enter groundwater and cause disease, e.g. cholera.
- Garbage attracts **rodents**, which can spread disease.
- Garbage creates an **eyesore**, which impacts negatively on tourism, especially **eco-tourism**.

Improper garbage disposal damages the environment

Pollution of marine and wetland ecosystems

- **Marine ecosystems** are **aquatic** ecosystems where the water contains dissolved compounds, especially salts, i.e. it is 'salty'. They include coral reefs, seagrass beds, rocky and sandy shores, mangrove swamps, estuaries and the open ocean.
- **Wetland ecosystems** are **transitional** ecosystems where dry land meets water and the water may be fresh, brackish or salt. They are areas of land that are covered with water for either part or all of the year, and are usually found alongside rivers, lakes and coastal areas. They include mangrove swamps, freshwater swamps, marshes and bogs.

A coral reef

A mangrove swamp

A seagrass bed

Because of their rich biodiversity and beauty, these ecosystems are major contributors to the **economies** of many small island developing states of the Caribbean through **tourism**, **fisheries** and **coastal protection**.

Many of these ecosystems are being **polluted** by untreated sewage, chemical fertilisers, pesticides, industrial waste, hot water, garbage and oil from oil spills. This pollution impacts negatively on both the overall **health** of the ecosystems and their **aesthetic appeal**. They are also being **overfished** and destroyed for **development** purposes, e.g. to build harbours or marinas.

Damage to **coral reefs**, **mangrove swamps** and other **marine and wetland ecosystems** results in a **loss** of:

- **Biodiversity**; coral reefs and mangrove swamps being some of the most biodiverse ecosystems on the planet.
- **Habitats** for many organisms, e.g. reef fish and mangrove oysters.
- **Natural resources**, e.g. fish, crabs, lobsters, oysters, seaweeds and wood.
- **Attractions** and **recreational sites** for tourists.
- **Nursery grounds** for reef fish which mangrove swamps provide, resulting in a reduction in population sizes of fish on reefs.
- **Nesting** and **breeding grounds** for birds, e.g. egret and scarlet ibis.
- **Protection** for shorelines against wave action and tidal forces, resulting in increased **coastal erosion.**
- **Flood control** provided by wetlands.

Due to the pollution and destruction of marine and wetland ecosystems and the coastal erosion that often follows, the **tourism** and **fishing industries** of Caribbean states are in danger of declining and this will have a negative impact on their economies.

The greenhouse effect and global warming

Carbon dioxide, water vapour, dinitrogen monoxide (N_2O) and methane (CH_4) are **greenhouse gases**. They form a layer around the Earth that lets radiation from the sun pass through but prevents much of it being reflected back into space. This radiation causes warming of the Earth which is known as the **greenhouse effect**.

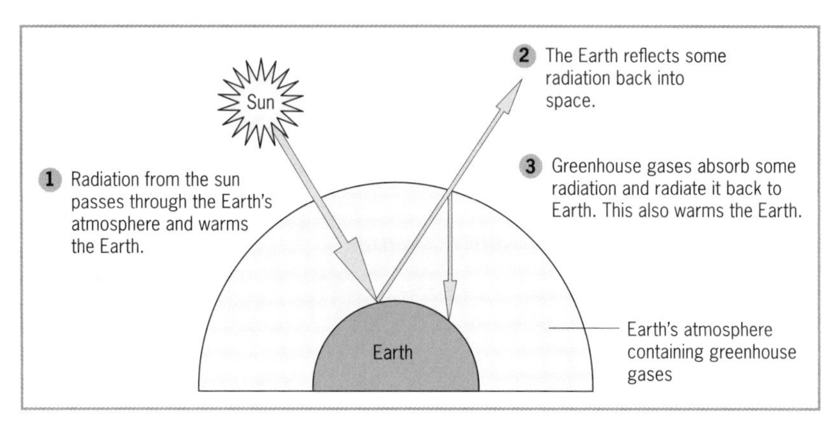

Figure 4.1 *The greenhouse effect*

An **increase** in greenhouse gases, especially **carbon dioxide**, caused by human activities, e.g. burning fossil fuels and deforestation, is **enhancing** the greenhouse effect and resulting in the Earth getting warmer. This warming, called **global warming**, is leading to **global climate change**. It is starting to cause:

- **Melting** of polar ice caps and glaciers.
- A **rise** in sea levels.
- **Flooding** of low-lying coastal areas.
- Changes in **global weather** patterns, e.g. some areas are becoming drier and others wetter than normal, while some areas are becoming colder and others hotter than normal.
- More **severe** weather events and natural disasters, e.g. colder winters, hotter summers, more extensive and frequent floods, droughts and wildfires, and more powerful hurricanes and tornadoes.
- Changes in **ecosystems** as the numbers of some species decline while numbers of other species increase, as some organisms immigrate into ecosystems while others emigrate out, and some species become extinct.
- Certain **diseases** to become more widespread, e.g. malaria.
- A rise in sea temperatures which can cause corals to **bleach** and **die**.

The **small island states** of the Caribbean are extremely **vulnerable** to the effects of global warming due to their small size, fragile ecosystems, low-lying coasts, vulnerability to natural disasters, constraints on transportation and communication and, in some islands, limited fresh water supplies.

Ocean acidification

Some carbon dioxide is absorbed by oceans causing the pH of the water to **decrease**, known as **ocean acidification**. This is expected to affect the ability of shellfish to produce and maintain their shells and of reef-building corals to produce their skeletons. This will reduce the chances of survival of these organisms which will have a negative impact on marine food chains and fishing industries, and lead to the erosion of coral reefs.

Conservation and restoration of the environment

It is important that the environment is **conserved** and **restored**. This can be achieved in a variety of ways.

Conserve and restore natural resources

- Use alternative energy sources, e.g. solar, wind and geothermal, instead of fossil fuels.
- Replace renewable resources, e.g. practise reforestation.
- Recycle resources, e.g. glass, plastic, metals and paper.
- Re-use materials, e.g. glass bottles.
- Use materials made from renewable resources instead of non-renewable resources, e.g. use cotton instead of synthetic fabrics such as polyester and nylon.
- Reduce soil erosion, e.g. never leave soil barren and prevent overgrazing by animals.
- Re-use land used in mining and landfills, e.g. replace soil and replant vegetation.
- Impose closed seasons and restrict catch sizes for overfished species, e.g. lobsters and sea eggs.
- Set up breeding and aquaculture programmes for endangered and overexploited species, e.g. captive breeding programmes and fish farms.
- Set up nature reserves, national parks and marine sanctuaries.
- Put legislation in place to make it illegal to kill endangered species, e.g. turtles.

Reduce pollution

- Use alternative energy sources that do not cause pollution instead of burning of fossil fuels.
- Use organic fertilisers instead of inorganic chemical fertilisers.
- Use natural, biodegradable pesticides and herbicides or biological control instead of synthetic pesticides and herbicides.
- Dispose of waste using appropriate methods and methods that produce harmless or useful end products:
    - Purify all effluent from factories.
    - Treat all sewage in sewage treatment plants and use the sludge as fertiliser and the water to irrigate crops.
    - Collect and recycle or re-use all recyclable waste, e.g. glass, plastic, metals and paper.
    - Compost all waste of plant origin, e.g. vegetable peelings and crop residues.
    - Use farmyard waste and waste from food industries to produce biogas.
- Use aerosol propellants and refrigerants that do not contain harmful chlorofluorocarbons (CFCs).
- Ensure gaseous emissions from factories are cleaned before they enter the environment.

Other strategies

- Develop **educational programmes** for people of all ages.
- Implement **monitoring programmes** to continually assess the health of ecosystems.
- Practise **organic agriculture** which involves using natural pesticides, herbicides and fertilisers, rotating crops and livestock, recycling organic matter back into the soil, practising soil conservation and using preventative disease control measures.
- Sign **international agreements** to control pollution and conserve natural resources.
- Pass **legislation** to protect the environment.

The growth and survival of populations

The rate at which a population **grows** depends on four factors: the **birth rate**, the **death rate**, the rate of movement of organisms **into** the population, i.e. immigration, and the rate of movement of organisms **out of** the population, i.e. emigration. The growth of a population may be represented by a **sigmoid growth curve** (see Figure 13.1, page 114).

When members of a species first **colonise** an area and start to reproduce, the birth rate exceeds the death rate and the population **increases** in size until the area cannot support any more individuals. This area is said to have reached its **carrying capacity**. At this point certain factors **limit** further population growth:

- **Food shortages** start to occur.
- **Overcrowding** begins to occur which causes increased **competition** for space, a mate, food and shelter in animals, and light, water and mineral salts in plants.
- **Diseases** start to spread more rapidly.
- **Predators** begin to increase in number.

These **limiting factors** cause the death rate to increase and the size of the population stabilises at a particular level, i.e. the **stationary phase** of growth is reached, where the birth and death rates are equal. If the death rate exceeds the birth rate, the population numbers **decrease**.

Population sizes can also be considerably **reduced** by:

- **Natural disasters** such as hurricanes, storms, floods, droughts, earthquakes, tsunamis and volcanic eruptions.
- **Invasive species**, i.e. species that colonise ecosystems in which they, themselves, are non-natives.
- **Pests**.

If environmental conditions become favourable again, surviving members of populations begin to reproduce and population numbers increase once more.

Human population growth

The **human population** is currently in the **phase of rapid growth**. It is growing at about 1.2% per year and has grown from 1 billion in 1804 to 7 billion in 2011, and is expected to reach about 8 billion by 2024.

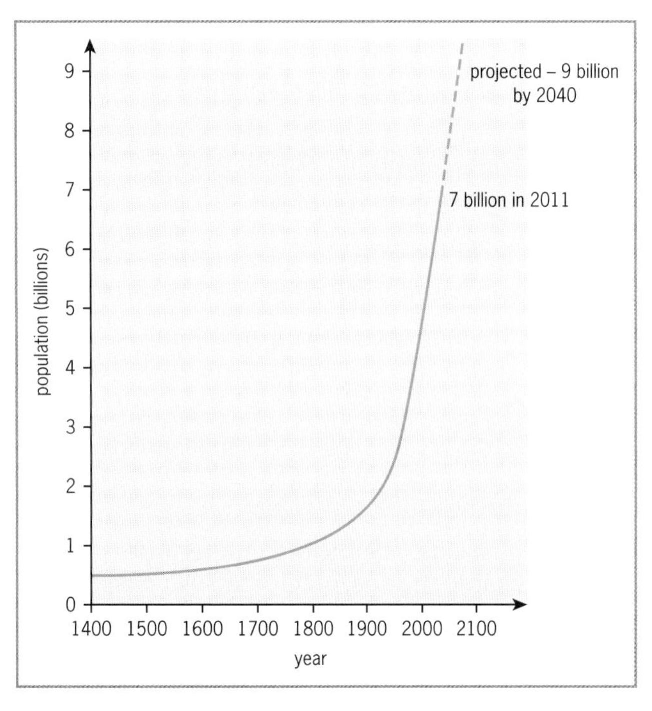

Figure 4.2 *Growth of the human population*

While humans are subjected to most of the **same constraints** as other organisms, the human population has grown exponentially since the mid-1800s because humans have:

- Developed **modern medicine** which has reduced the death rate from disease, improved infant survival and increased life expectancy.
- Improved **water supplies**, **sanitation** and **housing** which have reduced the death rate from disease.
- Improved **agricultural techniques** which have increased food production.
- Developed a better **nutritional understanding**, which has improved health and increased life expectancy.

Revision questions

1. By reference to specific examples, differentiate between renewable and non-renewable resources and explain how human activities are impacting on these resources.
2. How can the overuse of chemical fertilisers in agriculture affect aquatic environments?
3. Outline some of the harmful effects of the following pollutants which are produced by burning fossil fuels in industry:

   a sulfur dioxide **b** carbon monoxide **c** smoke.
4. Explain how the improper disposal of garbage is harmful to human health and the environment.
5. Discuss THREE reasons why marine and wetland ecosystems are of the upmost importance to the economies of many of the small island states of the Caribbean.
6. Name THREE greenhouse gases and explain the greenhouse effect.
7. An increase in greenhouse gases in the atmosphere is leading to global warming and global climate change. Outline FOUR consequences of global warming.
8. Outline FIVE different measures that should be taken to help conserve and restore our environment.
9. Suggest FOUR factors that affect the growth and survival of populations.

Exam-style questions – Chapters 1 to 4

Structured questions

1 The vegetation of a coastal ecosystem changes with distance from the sea, becoming progressively denser with plants growing larger as indicated in Figure 1 below.

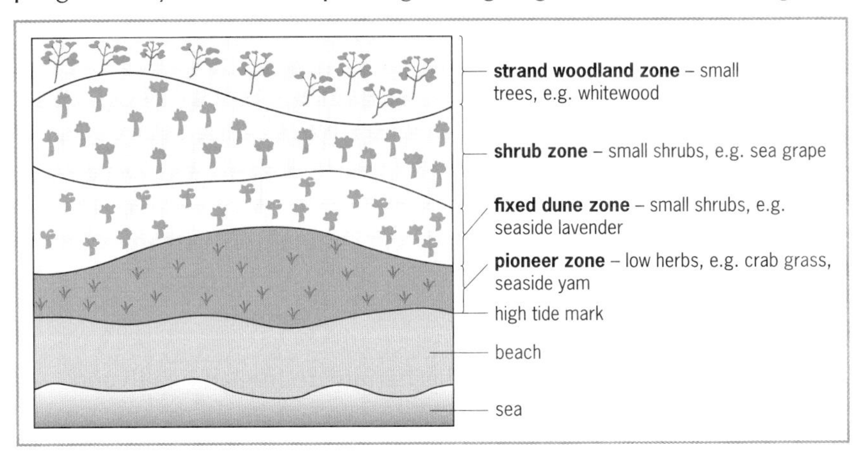

Figure 1 *A coastal ecosystem*

A Biology class went on a field trip to study a coastal ecosystem similar to that represented in the diagram.

a) i) Name a sampling technique that the students could use to get an idea of the distribution of plant species across the ecosystem. **(1 mark)**

ii) Explain how the students would use the technique named in **i)** above. **(3 marks)**

iii) Identify ONE abiotic factor that could be responsible for the change in vegetation across the ecosystem. **(1 mark)**

b) One group of students in the class used a 1 m^2 quadrat to determine the species density and species frequency of crab grass, turtle grass and seaside yam in the pioneer zone of the ecosystem. Their results are shown in Table 1 below.

Table 1 *Results from five quadrats*

Species	Number of organisms in each quadrat (Q)					Species density/ number of individuals per m^2	Species frequency/%
	Q1	Q2	Q3	Q4	Q5		
crab grass	0	5	8	3	9		
turtle grass	4	3	0	3	0		
seaside yam	12	9	11	13	10		

i) Outline how the students used the quadrat to obtain these results. **(2 marks)**

ii) Complete Table 1 to give the species density and species frequency of EACH species. **(2 marks)**

iii) Construct a bar chart to show the species density of the three species. **(3 marks)**

iv) Seaside yam is a succulent herb. Suggest ONE reason for its presence in the pioneer zone. **(2 marks)**

c) Another group of students from the class wishes to study the crawling animals present in the ecosystem. What piece of apparatus would they use? **(1 mark)**

Total 15 marks

2 a) Table 2 below shows the food sources of several organisms found in a woodland community.

Table 2 *Food sources of organisms found in a woodland community*

Organism	Food source
Slug	Leaves
Caterpillar	Leaves
Earthworm	Leaf litter
Frog	Slugs and caterpillars
Blackbird	Caterpillars and earthworms
Grass snake	Frogs
Mongoose	Grass snakes and blackbirds

i) Using only the information contained in Table 2, construct a food web for the woodland community. **(3 marks)**

ii) Identify from the food web in **i)** above, ONE predator and ONE organism that is likely to be its prey. **(2 marks)**

iii) What is the importance of predator/prey relationships within ecosystems? **(1 mark)**

iv) Suggest TWO consequences that might occur if a disease developed within the blackbird population causing most of its members to die. **(2 marks)**

b) Explain why the number of organisms decreases at successive trophic levels in a food chain. **(3 marks)**

c) Some organisms have special relationships with organisms of a different species. Name the organism that shares a special relationship with a leguminous plant, identify the type of relationship they share and state ONE way in which EACH organism benefits from the relationship. **(4 marks)**

Total 15 marks

Extended response questions

3 a) Using a labelled diagram, explain how carbon is recycled in nature. **(5 marks)**

b) Human activities are affecting the natural cycling of carbon.

i) Suggest TWO ways in which human activities are contributing to the increasing levels of atmospheric carbon dioxide. **(2 marks)**

ii) Outline THREE consequences of global warming, which is being caused by this increase in atmospheric carbon dioxide levels. **(3 marks)**

iii) Suggest TWO methods that can be employed to control atmospheric carbon dioxide levels. **(2 marks)**

c) It has been said that humans are "nature's curse". Give THREE different reasons you could put forward to support this statement, other than the information already used when answering part **b)** above. **(3 marks)**

Total 15 marks

4 a) Discuss THREE factors that have contributed to the dramatic increase in human population growth over the past 150 years. **(6 marks)**

b) Growth of the human population is having a negative effect on ecosystems worldwide, including the marine and wetland ecosystems of the Caribbean.

i) By reference to specific examples, distinguish between a marine ecosystem and a wetland ecosystem. **(4 marks)**

ii) Outline TWO ways in which marine and wetland ecosystems are being destroyed in the Caribbean. **(2 marks)**

iii) Discuss THREE consequences of this destruction to the countries of the Caribbean. **(3 marks)**

Total 15 marks

5 Cells

The **cell** is the basic structural and functional unit of living organisms. Some organisms are **unicellular**, being composed of a single cell; others are **multicellular**, being composed of many cells. Cells are so small that they can only be seen with a microscope and not with the naked eye.

Plant and animal cells

All plant and animal cells contain structures called **organelles** which are specialised to carry out one or more vital functions, e.g. the nucleus, mitochondria, chloroplasts and vacuoles. Organelles are found within the **cytoplasm** of the cells and most are surrounded by one or two **membranes**.

The following structures are found in **all** plant and animal cells:

- a **cell membrane** or **plasma membrane**
- **cytoplasm**
- a **nucleus**
- **mitochondria** (singular mitochondrion).

In addition to the above, **plant cells** also possess:

- a **cell wall**
- **chloroplasts**
- a large **vacuole**.

The cytoplasm and nucleus together are referred to as **protoplasm**.

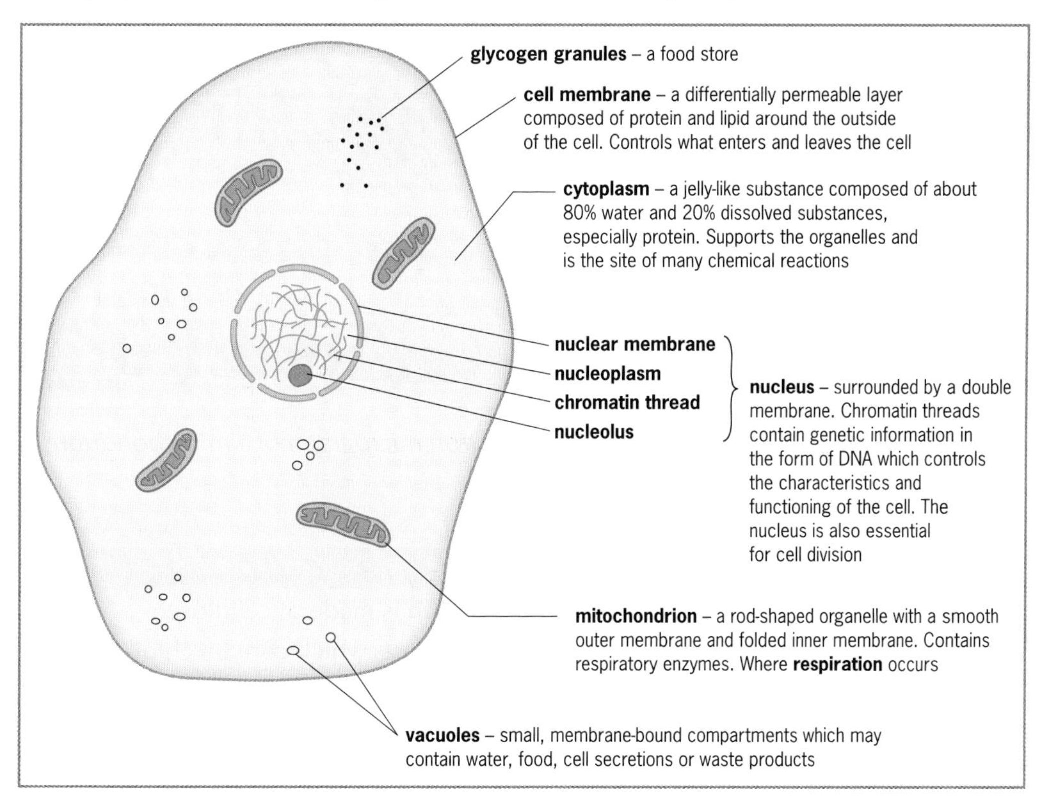

Figure 5.1 *Structure and function of the parts of a generalised animal cell*

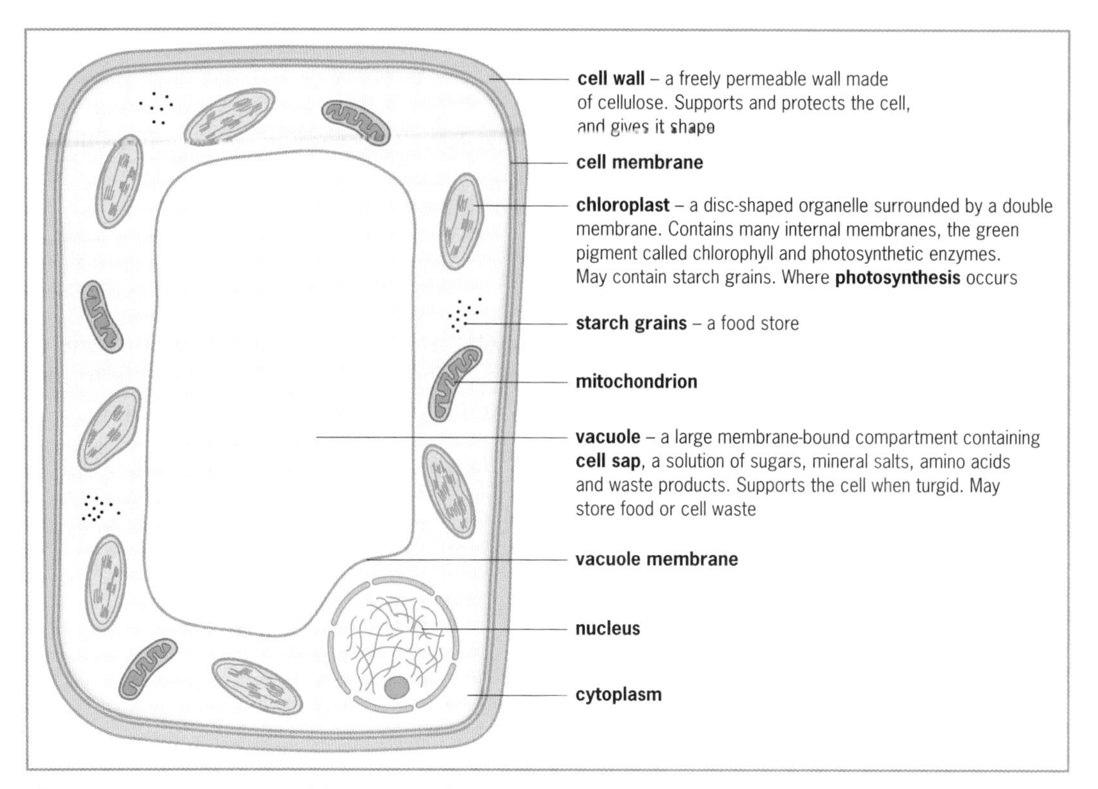

Figure 5.2 *Structure and function of the parts of a generalised plant cell*

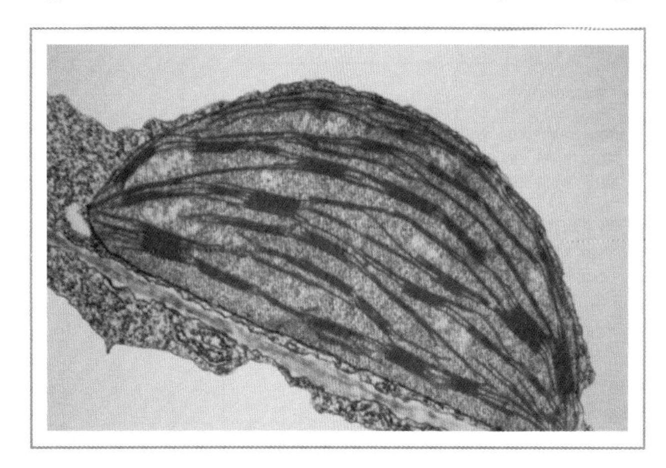

An electron micrograph of a chloroplast

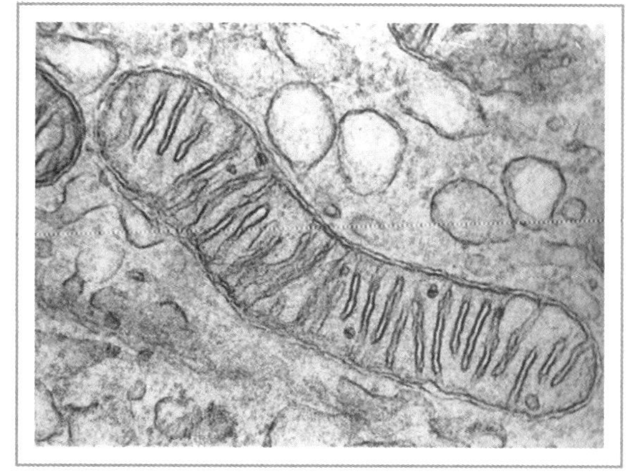

An electron micrograph of a mitochondrion

Table 5.1 *Plant and animal cells compared*

Animal cells	Plant cells
Do not have a cell wall.	Have a **cell wall** which is made of cellulose.
Do not have chloroplasts or chlorophyll.	Usually have **chloroplasts** which contain chlorophyll.
When present, the vacuoles are small and scattered throughout the cytoplasm and their contents vary.	Usually have one large, central **vacuole** which contains **cell sap**.
May contain **glycogen granules** as a food store.	May contain **starch grains** as a food store.
Can have a great variety of different shapes.	Have a regular shape, usually round, square or rectangular.

Microbe cells

Microbes or **micro-organisms** are extremely small organisms that include all members of the **Prokaryotae** kingdom, e.g. bacteria, many members of the **Protoctista** kingdom, e.g. amoeba, and some members of the **Fungi** kingdom, e.g. yeast.

The cells of **prokaryotes** lack a true nucleus and other membrane-bound organelles. Their DNA exists in a region called the **nucleoid**, which lacks a nuclear membrane, and also in smaller regions called **plasmids**.

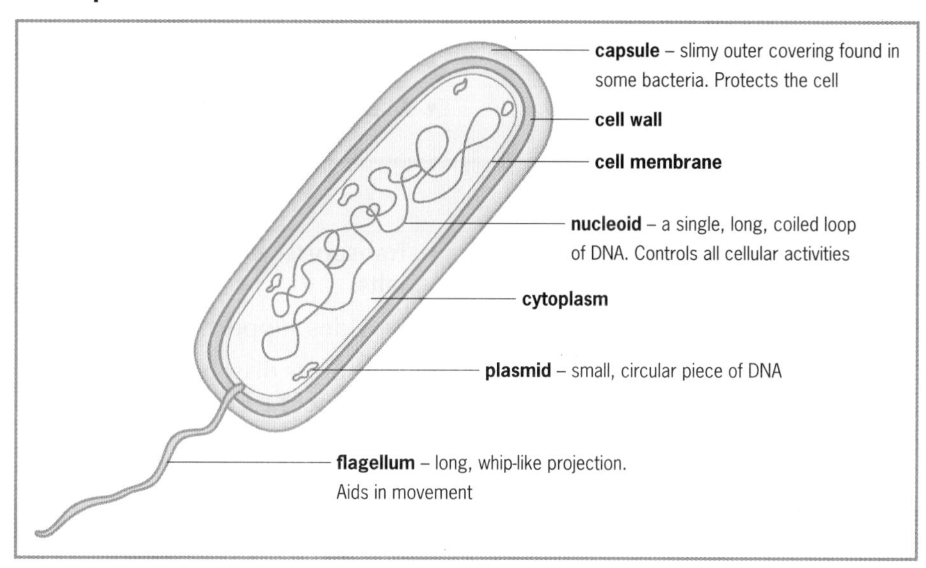

Figure 5.3 *A generalised bacterial cell*

The cells of **protoctists** and **fungi** all have true nuclei surrounded by nuclear membranes, and other membrane-bound organelles.

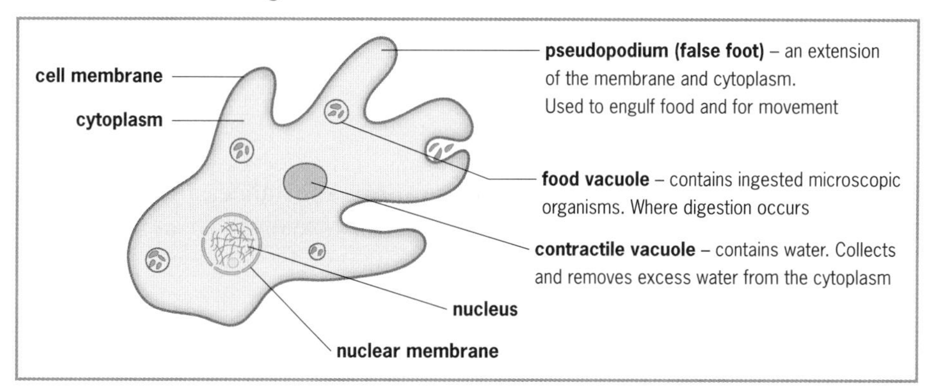

Figure 5.4 *An amoeba*

Cell specialisation

The body of a **unicellular organism** consists of **one unspecialised cell**. This cell can carry out **all** essential life processes.

The body of a **multicellular organism** is composed of **many cells**. To enable multicellular organisms to carry out all essential life processes **efficiently**, cells in their bodies become **specialised (differentiated)** to carry out specific functions, e.g. muscle cells in animals contract to bring about movement, mesophyll cells in leaves carry out photosynthesis. By becoming specialised, cells are **better able** to carry out their specific functions. Cells specialised to carry out a particular function then work together in groups called **tissues**. Tissues may contain one or, in some cases, more than one type of cell.

Table 5.2 *Some examples of plant tissues*

Name of tissue	What it is composed of	Where it is found	Functions
Epidermal tissue	Sheets of flattened epidermal cells.	Around the outside of leaves, young stems and roots.	• Protects the surfaces of leaves, stems and roots.
Packing tissue	Round or rectangular cells with large vacuoles called parenchyma cells.	Inside stems and roots.	• Fills spaces in stems and roots. • Supports non-woody plants when turgid. • Stores food.
Photosynthetic tissue	Round or rectangular cells containing chloroplasts called mesophyll cells.	Mainly in leaves.	• Makes food by photosynthesis.
Vascular tissue	Long tubes called xylem vessels and phloem sieve tubes with companion cells.	In leaves, stems and roots.	• Transports water and mineral salts. • Provides support. • Transports dissolved food substances.

Table 5.3 *Some examples of animal tissues*

Name of tissue	What it is composed of	Functions
Nerve tissue	Nerve cells or neurones.	• Conducts nerve impulses.
Muscle tissue	Muscle cells.	• Brings about movement on contraction.
Epithelial tissue	Sheets of cells.	• Covers and often protects inner and outer surfaces of the body, e.g. lines blood vessels and forms the outer layers of the skin.
Connective tissue	A variety of cells surrounded by extracellular material. Examples:	
	Blood tissue – red blood cells, white blood cells and platelets surrounded by plasma.	• Transports various substances around the body, e.g. food and oxygen. • Helps fight disease.
	Adipose (fat) tissue – fat cells surrounded by extracellular material.	• Insulates the body. • Serves as a food reserve. • Protects the body by acting as 'padding'.

Different tissues are then grouped together to form specialised **organs** which may perform one or more functions, e.g. the **skin** is composed of epithelial, connective and nerve tissues; the **leaves** of plants are composed of epidermal, photosynthetic and vascular tissues.

Organs work together in **organ systems**, e.g. the digestive, nervous and blood vascular systems in animals, and the transpiration and translocation systems in plants. Systems then work together in an organised way to form a **multicellular organism**.

i.e. cells ⟶ tissues ⟶ organs ⟶ organ systems ⟶ a multicellular organism

Revision questions

1. By means of a fully labelled and annotated diagram, describe the structure of a generalised animal cell.
2. State the function of any THREE of the structures you labelled in question **1**.
3. Give FOUR differences between the structure of an animal cell and that of a plant cell.
4. What features would enable a scientist to distinguish a bacterial cell from other cells when viewed under a microscope?
5. Why does cell specialisation occur in large multicellular organisms but not in small unicellular organisms?
6. What is a tissue?
7. Name TWO different tissues found in animals and TWO different tissues found in plants and give the functions of EACH.

Movement of substances into and out of cells

Substances can move into and out of cells, and from cell to cell, by **three** different processes:

- **diffusion**
- **osmosis**
- **active transport**.

Diffusion

***Diffusion** is the net movement of particles from an area of higher concentration to an area of lower concentration until the particles are evenly distributed.*

The particles are said to move **down a concentration gradient**. Particles in gases, liquids and solutions are capable of diffusing. Diffusion is the way cells obtain many of their requirements and get rid of their waste products which, if not removed, would poison them.

The importance of diffusion in living organisms

- **Oxygen**, for use in **aerobic respiration**, moves into organisms through gaseous exchange surfaces and into cells by diffusion.
- **Carbon dioxide**, produced in **aerobic respiration**, moves out of cells and out of organisms through gaseous exchange surfaces by diffusion.
- **Carbon dioxide**, for use in **photosynthesis**, moves into leaves and plant cells by diffusion.
- **Oxygen**, produced in **photosynthesis**, moves out of plant cells and leaves by diffusion.
- Some of the **glucose** and **amino acids** produced in **digestion** are absorbed through the cells in the ileum and capillary walls and into the blood by diffusion.

Osmosis

Osmosis is a special form of diffusion.

*Osmosis is the movement of **water molecules** through a differentially permeable membrane from a solution containing a lot of water molecules, e.g. a dilute solution (or water), to a solution containing fewer water molecules, e.g. a concentrated solution.*

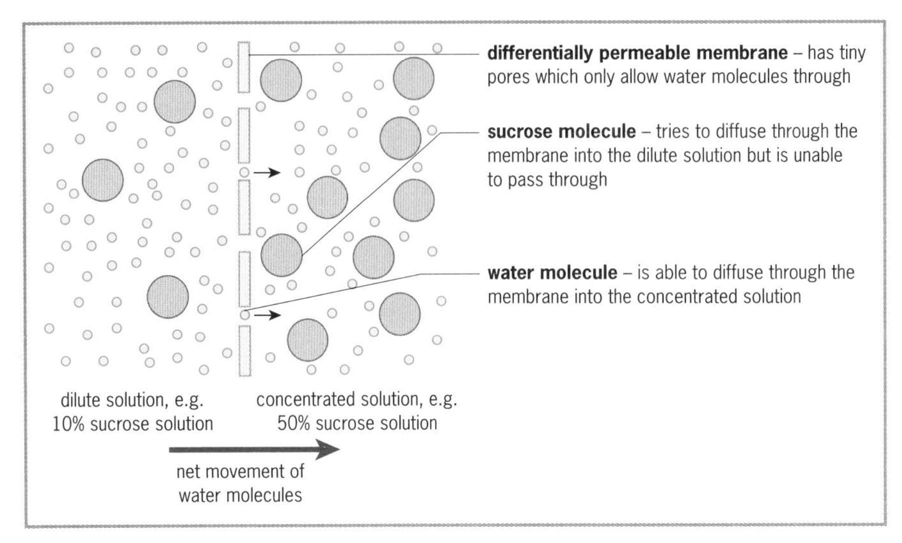

Figure 5.5 *Explanation of osmosis*

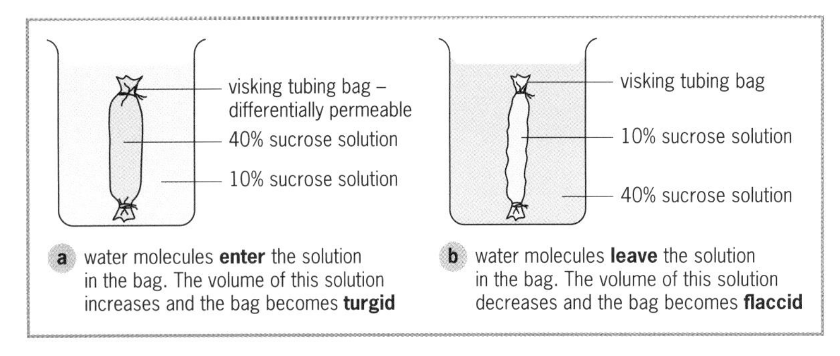

Figure 5.6 *Demonstrating osmosis*

In any cell, the **cell membrane** is differentially permeable. There is always **cytoplasm**, a solution of protein and other substances in water, on the inside of the membrane and usually a solution on the outside. **Water molecules**, therefore, move into and out of cells by **osmosis**.

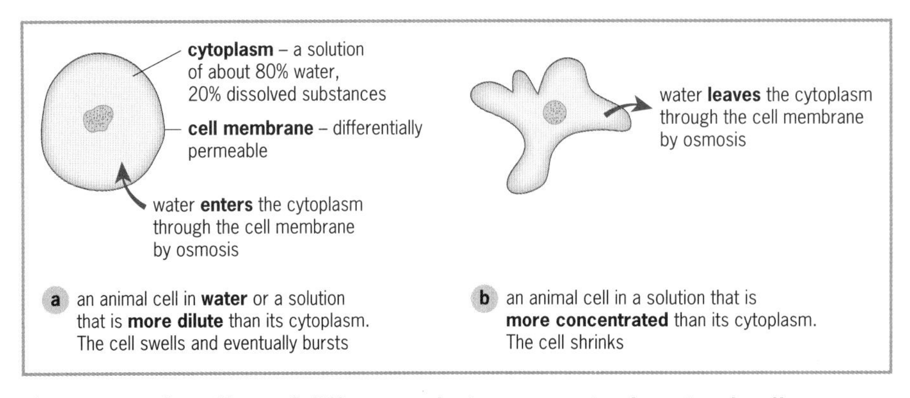

Figure 5.7 *The effect of different solutions on a single animal cell*

Plant cells are surrounded by strong, freely permeable **cell walls**. Because of this, they behave differently from animal cells when placed in different solutions.

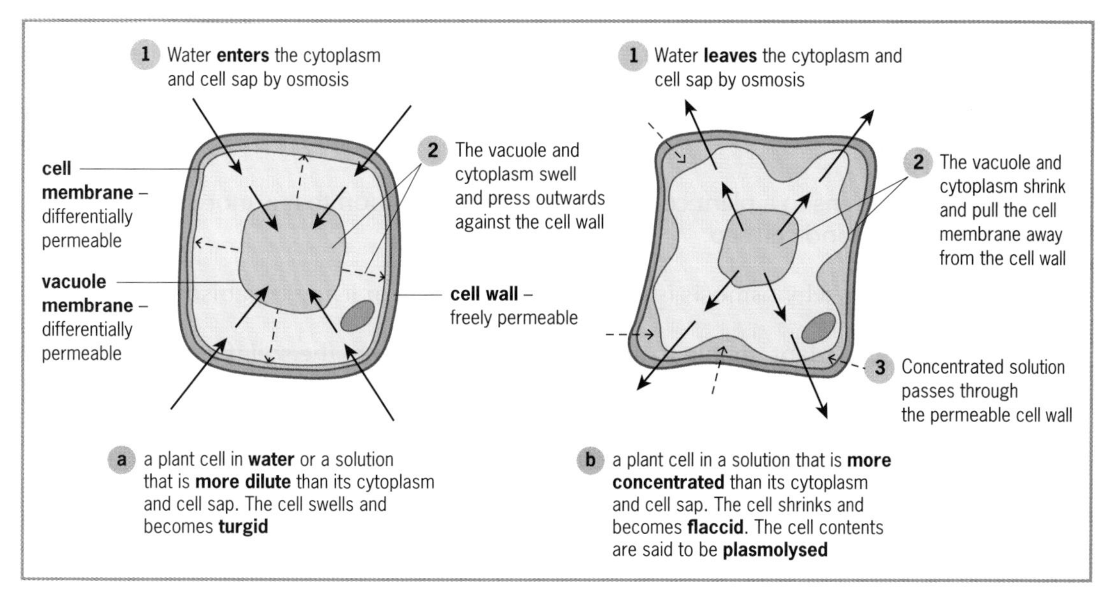

Figure 5.8 *The effect of different solutions on a single plant cell*

The importance of osmosis in living organisms

- All cells are kept **hydrated** by water moving into them by osmosis.
- Plant cells are kept **turgid** by water moving into them by osmosis. This causes non-woody stems to stand upright and keeps leaves firm.
- Water is kept **moving** through plants by osmosis occurring in the cells of roots and leaves. This ensures that leaves get water for photosynthesis.
- The **size** of stomatal pores is regulated by osmosis occurring in the guard cells. This controls the loss of water from the leaves of plants.
- Water is **reabsorbed** into the blood from the filtrate in the kidney tubules by osmosis. This prevents the body from losing too much water.

Active transport

During **active transport**, particles move through membranes **against** a concentration gradient. **Energy** produced in respiration is used to move the particles through the membranes from areas of lower concentration to areas of higher concentration. Active transport allows cells to accumulate high concentrations of important substances, e.g. glucose, amino acids and ions.

The importance of active transport in living organisms

- **Mineral ions** move from the soil into plant roots by active transport.
- **Sugars** produced in photosynthesis move into the phloem in leaves by active transport.
- Some of the **glucose** and **amino acids** produced in digestion are absorbed from the ileum into the blood by active transport.
- **Useful substances** are reabsorbed from the filtrate in the kidney tubules into the blood by active transport.

Revision questions

8. Define the term 'diffusion'.

9. Cite FOUR reasons to support the fact that diffusion is important to living organisms.

10. What is osmosis?

11. Explain what happens to a plant cell if it is placed in a solution that is more dilute than its cytoplasm and cell sap.

12. Give FOUR reasons why osmosis is important in the lives of living organisms.

13. Why is the root of a plant unable to absorb mineral salts from the soil if it is given a poison that prevents respiration?

6 The chemistry of living organisms

Living organisms are composed of about 22 different **chemical elements.** These are combined to form a great variety of **compounds**. Six major elements make up almost 99% of the mass of the human body, as shown in Figure 6.1.

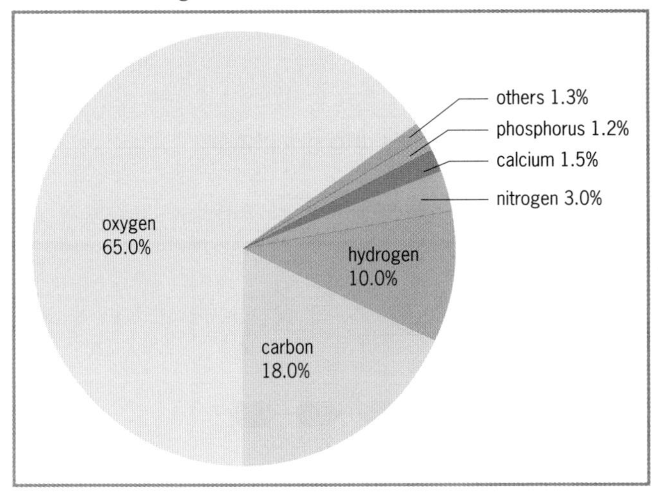

Figure 6.1 *The percentage, by mass, of the major elements that make up the human body*

The main **compounds** formed from these elements include water, an **inorganic compound** that makes up about 65% of the body, and carbohydrates, proteins and lipids which are all **organic compounds.**

Carbohydrates

Carbohydrates include sugars and starches. They are **molecules** composed of carbon, hydrogen and oxygen atoms. The ratio of hydrogen atoms to oxygen atoms is always 2:1. The simplest carbohydrate molecule has the formula $C_6H_{12}O_6$. Carbohydrates can be classified into **three** groups: **monosaccharides**, **disaccharides** and **polysaccharides**.

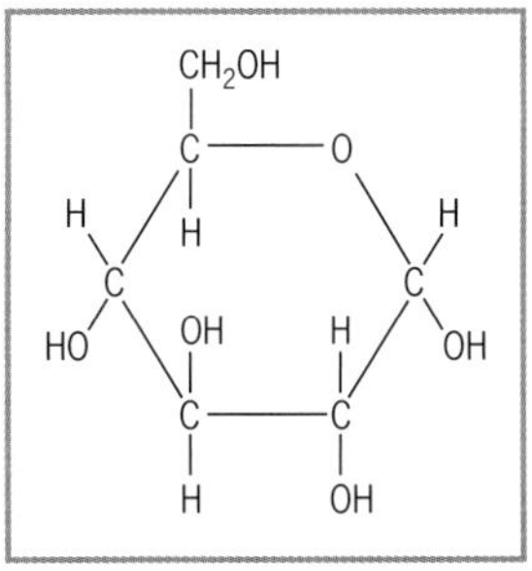

Figure 6.2 *A glucose molecule*

Table 6.1 *The three groups of carbohydrates*

Group	Properties	Formula	Examples
Monosaccharides (single sugars)	Have a sweet taste. Soluble in water.	$C_6H_{12}O_6$	Glucose Fructose Galactose
Disaccharides (double sugars)	Have a sweet taste. Soluble in water.	$C_{12}H_{22}O_{11}$	Maltose Sucrose Lactose
Polysaccharides (complex carbohydrates)	Do not have a sweet taste. Insoluble in water.	$(C_6H_{10}O_5)_n$	Starch Cellulose Glycogen (animal starch)

Disaccharides are formed by chemically joining two monosaccharide molecules together with the loss of a water molecule from between, a process called **condensation** or **dehydration synthesis**:

$$C_6H_{12}O_6 + C_6H_{12}O_6 \longrightarrow C_{12}H_{22}O_{11} + H_2O$$

glucose + glucose ⟶ maltose

glucose + fructose ⟶ sucrose

glucose + galactose ⟶ lactose

Polysaccharides are formed by the **condensation** of many monosaccharides into straight or branched chains.

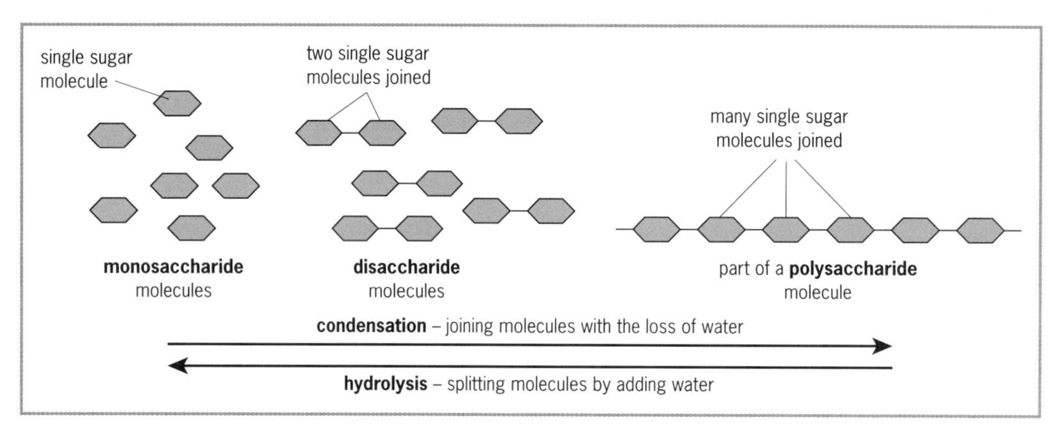

Figure 6.3 *The three types of carbohydrates*

Lipids

Lipids are fats and oils. They feel greasy and are insoluble in water. Lipids are **molecules** composed of carbon, hydrogen and oxygen atoms. Their molecules have fewer oxygen atoms than carbohydrate molecules, e.g. beef fat has the formula $\mathbf{C_{57}H_{110}O_6}$. Each lipid molecule is composed of **four** smaller molecules joined together; three **fatty acid** molecules and one **glycerol** molecule.

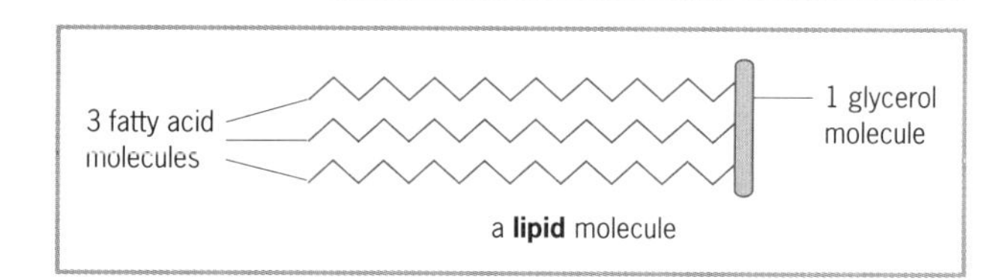

Figure 6.4 *A lipid molecule*

Proteins

Proteins are **molecules** composed of carbon, hydrogen, oxygen, nitrogen, and sometimes sulfur and phosphorus atoms. These atoms form small molecules known as **amino acids**. There are 20 different common amino acids. Protein molecules are formed by the **condensation** of hundreds or thousands of amino acid molecules in long chains. The links between adjacent amino acid molecules are called **peptide links**. The chains then fold to give each type of protein molecule a specific shape.

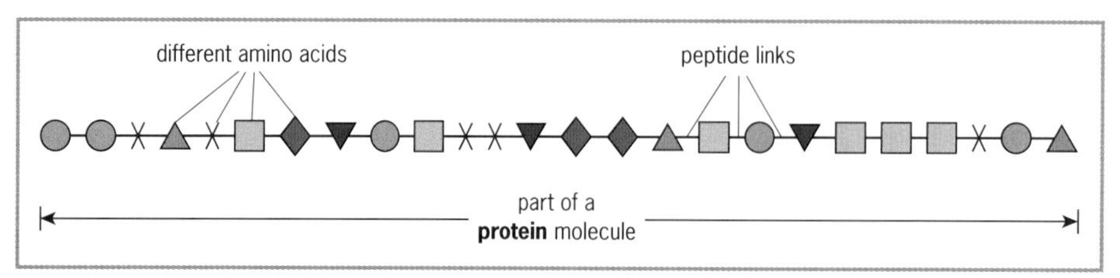

Figure 6.5 *Part of a protein molecule*

Some proteins are soluble in water, e.g. haemoglobin; others are insoluble, e.g. collagen.

Recognising carbohydrates, proteins and lipids

Tests can be performed in the laboratory to identify carbohydrates, proteins and lipids. Apart from the tests for lipids, the tests are usually carried out on about 2 cm^3 of a solution of the test substance in a test tube.

Table 6.2 *Laboratory tests to identify carbohydrates, proteins and lipids*

Food substance	Test	Positive result
Reducing sugars – monosaccharides and some disaccharides, e.g. glucose, maltose	Add an equal volume of **Benedict's solution** and shake. Heat the mixture.	An **orange-red** precipitate forms.
Non-reducing sugars – some disaccharides, e.g. sucrose	Add a few drops of dilute **hydrochloric acid** and heat for 1 minute. Add **sodium hydrogencarbonate** until effervescence stops. Add an equal volume of **Benedict's solution** and shake. Heat the mixture.	An **orange-red** precipitate forms. The acid **hydrolyses** the disaccharide molecules to monosaccharide molecules. The sodium hydrogencarbonate neutralises the acid allowing the Benedict's solution to react with the monosaccharides.
Starch	Add a few drops of **iodine solution** and shake.	Solution turns **blue-black**.
Protein – the biuret test	Add an equal volume of **sodium hydroxide solution** and shake. Add drops of dilute **copper sulfate solution** and shake. Or add an equal volume of **biuret reagent** and shake.	Solution turns **purple**.
Lipid – the emulsion test	Place 4 cm^3 of **ethanol** in a dry test tube. Add one drop of test substance and shake. Add an equal volume of **water** and shake.	A **milky-white** emulsion forms.
Lipid – the grease spot test	Rub a drop of test substance onto **absorbent paper**. Leave for 10 minutes.	A **translucent mark** remains.

Benedict's solution forms an orange-red precipitate with reducing sugars

Iodine solution turns starch blue-black

Enzymes

Enzymes are biological catalysts produced by all living cells. They speed up chemical reactions occurring in living organisms without being changed themselves.

Enzymes are **proteins** that living cells produce from amino acids obtained from the diet in animals, or manufactured in plants. Without enzymes, chemical reactions would occur too slowly to maintain life.

Examples:

- **Amylase** catalyses the breakdown of **starch** into sugars, mainly **maltose**. It is present in saliva, pancreatic juice and germinating seeds.

$$\text{starch} \xrightarrow{\text{amylase}} \text{maltose}$$

- **Catalase** catalyses the breakdown of hydrogen peroxide into water and oxygen:

$$\text{hydrogen peroxide} \xrightarrow{\text{catalase}} \text{water} + \text{oxygen}$$

or

$$2H_2O \xrightarrow{\text{catalase}} 2H_2O + O_2$$

Catalase is found in most cells. It prevents the build-up of harmful hydrogen peroxide which is produced as a by-product of many chemical reactions occurring in cells.

Properties of enzymes

All enzymes have similar **properties**:

- Enzymes are **specific**, i.e. each type of enzyme catalyses only one type of reaction.
- Enzymes work best at a particular temperature known as the **optimum temperature**. This is about 37 °C for human enzymes.

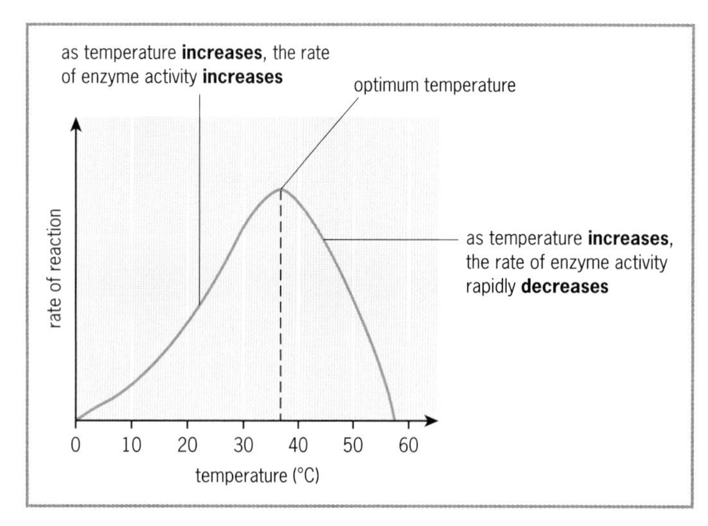

Figure 6.6 *The effect of temperature on the rate of a reaction catalysed by enzymes*

- High temperatures **denature** enzymes, i.e. the shape of the enzyme molecules changes so that they are inactivated. Enzymes start to be denatured at about 40 °C to 45 °C.

- Enzymes work best at a particular pH known as the **optimum pH**. This is about pH 7 for most enzymes.

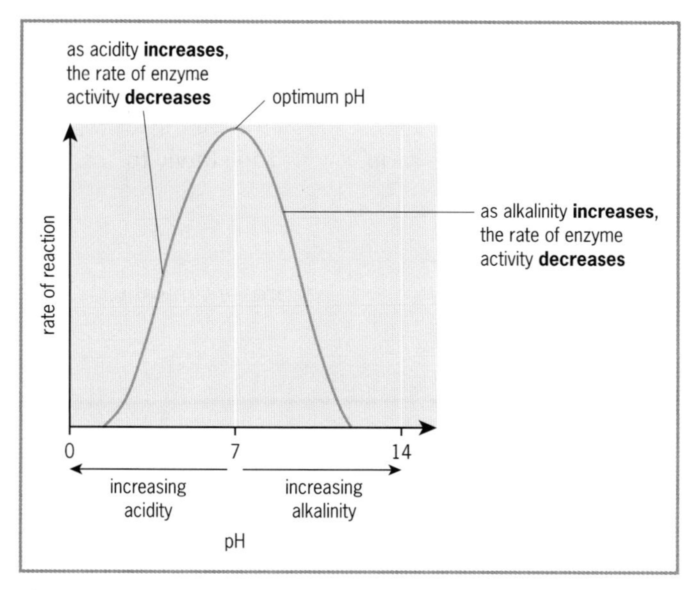

Figure 6.7 *The effect of pH on the rate of a reaction catalysed by enzymes*

- Extremes of acidity or alkalinity **denature** most enzymes.
- The action of enzymes is **helped** by certain vitamins and minerals, e.g. vitamin B_1 helps the action of respiratory enzymes.
- The action of enzymes is **inhibited** by certain poisons, e.g. arsenic and cyanide.

Revision questions

1. Describe the simple molecular structure of EACH of the following:
   a polysaccharides **b** proteins **c** lipids.
2. Distinguish between condensation and hydrolysis.
3. You are given three solutions labelled X, Y and Z and told that they contain starch, glucose and gelatin (a protein), respectively. Describe THREE laboratory tests you could perform to confirm what you are told about X, Y and Z.
4. What are enzymes?
5. Name the enzyme responsible for the breakdown of starch.
6. Explain the effect that temperature has on enzyme activity.
7. Other than the effect of temperature on enzyme activity, give THREE other properties of enzymes.

7 Nutrition

All living organisms need **food** to provide them with energy, and to enable them to grow and produce important chemicals for cellular processes.

***Nutrition** is the process by which living organisms obtain or make food.*

The types of nutrition

There are **two** types of nutrition:

- **autotrophic nutrition**
- **heterotrophic nutrition**.

Autotrophic nutrition

Autotrophic nutrition occurs in green plants and some bacteria. These organisms, called **autotrophs**, use simple inorganic compounds, e.g. carbon dioxide, water and minerals to **manufacture** complex organic food substances, e.g. carbohydrates, proteins, lipids and vitamins. Autotrophic nutrition requires a source of **energy**. The main type is **photosynthesis** which occurs in green plants and uses energy from **sunlight**.

Heterotrophic nutrition

Heterotrophic nutrition occurs in animals, fungi and most bacteria. These organisms, called **heterotrophs**, obtain **ready-made** organic food from their environment. There are **three** types:

- **Holozoic nutrition** occurs in most animals. The organisms obtain organic food by consuming other organisms. The complex organic food is **ingested** (taken in) by the organism and then **digested** (broken down) into simpler organic substances within the body of the organism.
- **Saprophytic nutrition** occurs in fungi and most bacteria. The organisms, called **saprophytes**, obtain organic food from the dead remains of other organisms. They digest the complex organic food **outside** their bodies and then **absorb** the simpler organic substances produced.
- **Parasitic nutrition** occurs in some plants, animals, fungi and bacteria. The organisms, called **parasites**, obtain organic food from the body of another living organism called the **host**. The host is usually harmed.

Photosynthesis in green plants

***Photosynthesis** is the process by which green plants convert carbon dioxide and water into glucose by using sunlight energy absorbed by chlorophyll in chloroplasts.*

Oxygen is produced as a by-product. The process can be summarised by the following **equation**:

$$\text{carbon dioxide} + \text{water} \xrightarrow[\text{by chlorophyll}]{\text{sunlight energy absorbed}} \text{glucose} + \text{oxygen}$$

or

$$6CO_2 + 6H_2O \xrightarrow[\text{by chlorophyll}]{\text{sunlight energy absorbed}} C_6H_{12}O_6 + 6O_2$$

Photosynthesis occurs in any plant structure that contains **chlorophyll**, i.e. is green; however, it mainly occurs in the **leaves**.

The two stages of photosynthesis

Photosynthesis occurs in the **chloroplasts** of plant cells, is catalysed by **enzymes** and occurs in **two** stages:

- **The light stage**

  The **light stage** or **light-dependent stage** requires **light energy**. The light energy is absorbed by the **chlorophyll** in chloroplasts and is used to split **water molecules** into **hydrogen** and **oxygen**. The oxygen (O_2) is a waste product and is released as a **gas**.

- **The dark stage**

  The **dark stage** or **light-independent stage** takes place whether or not light is present. The hydrogen atoms (H), produced in the light stage, reduce the **carbon dioxide molecules** forming **glucose**. The dark stage requires **enzymes**.

Conditions needed for photosynthesis

To take place, photosynthesis requires the following **six** conditions:

- **Carbon dioxide** which diffuses into the leaf from the air through the **stomata**.
- **Water** which is absorbed from the soil by the roots.
- **Sunlight energy** which is absorbed by the chlorophyll in chloroplasts.
- **Chlorophyll**, the green pigment that is present in chloroplasts.
- **Enzymes** which are present in chloroplasts.
- A **suitable temperature** between about 5 °C and 40 °C so that enzymes can function.

Certain **mineral ions** are also indirectly required since they are needed for plants to manufacture **chlorophyll**, e.g. magnesium (Mg^{2+}), iron (Fe^{3+}) and nitrate (NO_3^-) ions.

Adaptations of leaves for photosynthesis

Photosynthesis occurs mainly in the **leaves** of green plants. All leaves consist of a flat part called the **lamina** which is made up of several layers of cells. Photosynthesis takes place in the **mesophyll cells** of the lamina. The lamina is attached to the plant stem by the **petiole** or leaf stalk. **Vascular tissue** composed of xylem vessels, phloem sieve tubes and companion cells, runs through the petiole and throughout the lamina in the **midrib** and **veins** so that all the mesophyll cells are close to the vascular tissue.

Leaf structure is **adapted** both externally and internally to carry out photosynthesis as **efficiently** as possible.

External features of a leaf

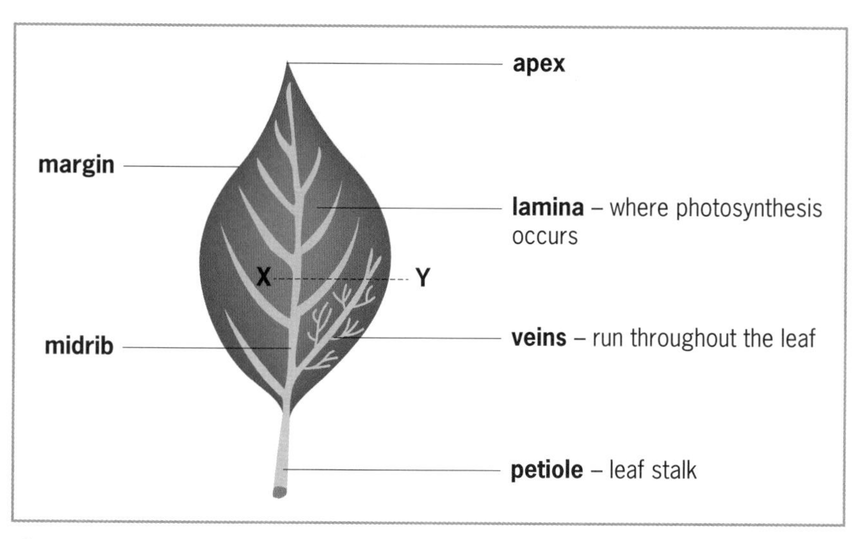

Figure 7.1 *The external features of a leaf*

- The lamina is usually **broad** and **flat**. This gives it a large surface area to absorb sunlight energy and carbon dioxide.
- The lamina is usually **thin**. This allows sunlight energy and carbon dioxide to reach all the cells.
- The lamina is **held out flat** by the **veins**. This maximises its exposure to the sunlight.
- The lamina usually lies at **90°** to the sunlight. This maximises its exposure to the sunlight.
- The laminae are **spaced out** around stems. This maximises each one's exposure to the sunlight.

The internal structure of a leaf

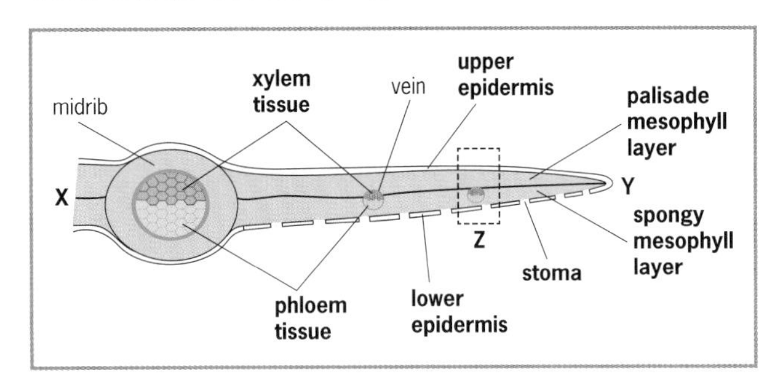

Figure 7.2 *A transverse section through part of a leaf lamina: X–Y in Figure 7.1*

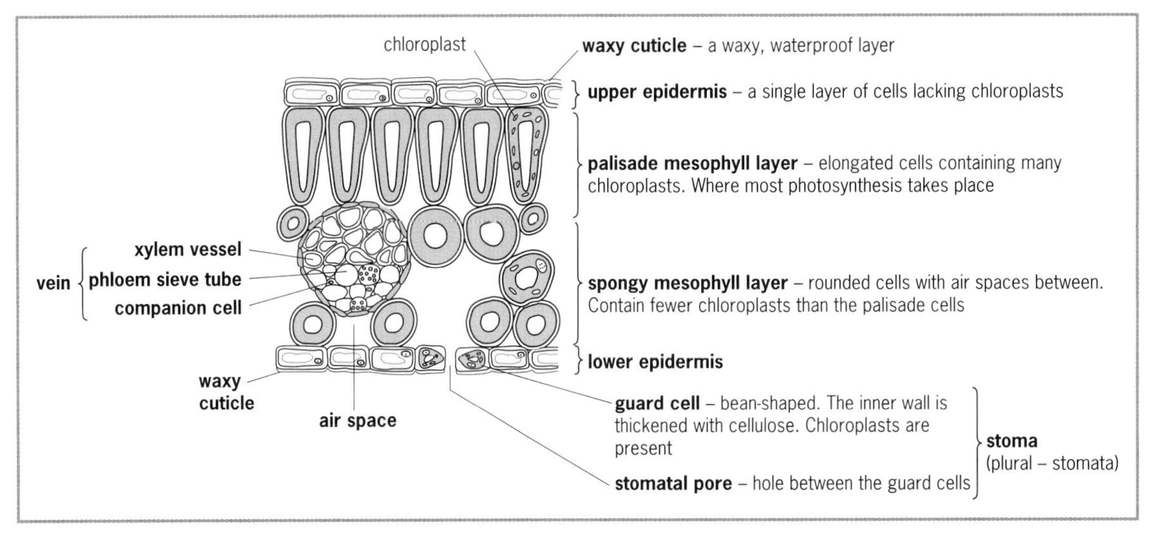

Figure 7.3 *The cellular structure of a leaf as seen in a transverse section: Z in Figure 7.2*

- **Waxy cuticles** on the outside of both the upper and lower epidermis are **waterproof** so they can prevent leaves losing water that is needed for photosynthesis.
- **Stomatal pores**, which are present throughout the lower epidermis, allow carbon dioxide to diffuse into the leaf and oxygen to diffuse out.
- The **palisade mesophyll cells**, which are directly below the upper epidermis and closest to the sunlight, contain a **large number of chloroplasts** to maximise the amount of light energy absorbed.
- The **palisade mesophyll cells** are arranged at **90°** to the leaf's surface to minimise the loss of sunlight energy which occurs as it passes through cell walls, and also to allow the chloroplasts to move to the top of the cells in dim light to maximise the amount of light absorbed.

Stomata in the lower epidermis of a leaf

- **Intercellular air spaces** between the **spongy mesophyll cells** allow carbon dioxide to diffuse to all the mesophyll cells and oxygen to diffuse away.
- **Xylem vessels** in the veins running throughout the leaf supply all the mesophyll cells with water and mineral ions.
- **Phloem sieve tubes** in the veins transport the soluble food made in photosynthesis away from the mesophyll cells to other parts of the plant.

Environmental factors that affect the rate of photosynthesis

Four main factors affect the rate of photosynthesis: light, carbon dioxide, temperature and water. The rate of photosynthesis is limited by which of these factors is in the **shortest supply**. This factor is known as the **limiting factor**.

- **Light** limits the rate between dusk and dawn, and also during the winter months in temperate climates.
- **Temperature** limits the rate during the winter months in temperate climates.
- **Water** limits the rate during the dry season in tropical climates and when the ground is frozen in temperate climates.
- **Carbon dioxide** limits the rate during the day in most climates since the concentration of carbon dioxide in the air is very low, i.e. about 0.04%.

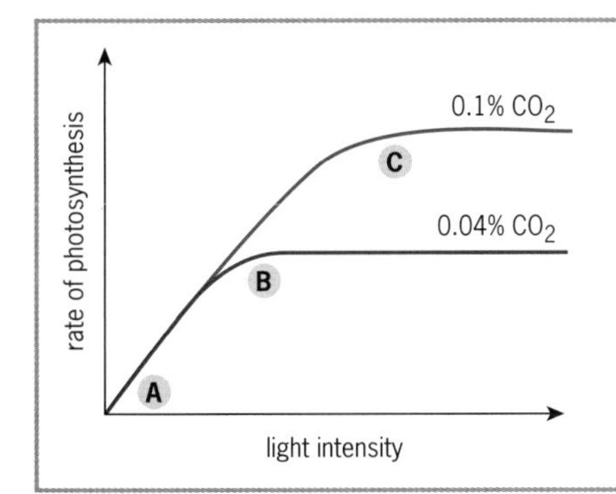

- Between points **A** and **B** light intensity is the limiting factor: as light intensity increases, the rate of photosynthesis increases.
- At point **B** some factor other than light limits the rate, e.g. carbon dioxide.
- If the carbon dioxide concentration is increased to 0.1%, the rate of photosynthesis increases as light intensity increases to point **C** where some factor other than light again limits the rate.

Figure 7.4 *The effect of light intensity and carbon dioxide concentration on the rate of photosynthesis*

The fate of glucose produced in photosynthesis

Several things can happen to the **glucose** made in photosynthesis:

- It can be used by the leaf cells in **respiration** to produce energy.
- It can be condensed to **starch** by the leaf cells and stored. The starch can then be hydrolysed back to glucose, e.g. during the night.
- It can be converted to other **organic substances** by leaf cells, e.g. amino acids and protein, vitamins or chlorophyll.
- It can be converted to **sucrose** and **transported**, via the phloem, to other parts of the plant such as growing parts and storage organs, where it can be converted to:
  - **Glucose** and used in **respiration** to produce energy.
  - **Cellulose** and used to make **cell walls** in growing parts.
  - **Starch** and **stored**.
  - **Amino acids** and **protein** by the addition of nitrogen from nitrates, and sulfur from sulfates obtained from the soil. Protein is then used for **growth**.
  - **Lipids** and **stored**, mainly in seeds.

Mineral nutrition in plants

Plants require a variety of **minerals** for healthy growth and development. These are absorbed from the soil by plant roots in the form of **ions**.

Table 7.1 *The major minerals required by plants*

Element	Form in which plants obtain the element	Functions	Results of deficiency
Nitrogen	Nitrate ions, NO_3^-	• To make proteins used for plant growth. • To make chlorophyll.	• Poor growth. • **Chlorosis** (yellowing) of leaves, especially older leaves. • Underdeveloped leaves.
Magnesium	Magnesium ions, Mg^{2+}	• To make chlorophyll; magnesium forms part of the chlorophyll molecule.	• **Chlorosis** of leaves.
Phosphorus	Phosphate ions, PO_4^{3-}	• To make ATP. • To make some proteins.	• Stunted growth, i.e. short stems. • Dull, purplish green leaves with curly brown edges. • Poor root growth.
Potassium	Potassium ions, K^+	• To help maintain the correct salt balance in cells. • To help in photosynthesis.	• Leaves have yellow-brown margins and brown spots that give a mottled appearance. • Premature death of leaves.
Sulfur	Sulfate ions, SO_4^{2-}	• To make proteins.	• Poor growth. • **Chlorosis** of leaves.
Calcium	Calcium ions, Ca^{2+}	• To make cell walls in the tips of growing roots and shoots.	• Poor, stunted growth. • Death of the growing tips of roots and shoots. • Poor bud development.

Revision questions

1. Distinguish between autotrophic nutrition and heterotrophic nutrition.
2. What happens when an organism feeds saprophytically?
3. Explain in detail how a green leaf manufactures glucose from carbon dioxide and water.
4. Write a balanced chemical equation to summarise the process of photosynthesis.
5. Describe FOUR ways in which the internal structure of a green leaf is adapted to carry out photosynthesis.
6. Identify THREE environmental factors that affect the rate at which green plants produce glucose.
7. Suggest THREE ways that plants can utilise the glucose produced in photosynthesis.
8. Why are plants growing in a soil deficient in nitrogen usually small with yellow leaves?

Heterotrophic nutrition in humans

Heterotrophic nutrition in humans involves the following **five** processes:

- **Ingestion**: the process by which food is **taken into** the body via the mouth.
- **Digestion**: the process by which food is **broken down** into simple, soluble food molecules.
- **Absorption**: the process by which the soluble food molecules, produced in digestion, **move into** the body fluids and body cells.
- **Assimilation**: the process by which the body **uses** the soluble food molecules absorbed after digestion.
- **Egestion** or **defaecation**: the process by which undigested food material is **removed** from the body.

A balanced diet

The **food** an animal eats is called its **diet**. Humans must consume a **balanced diet** each day. This must contain carbohydrates, proteins, lipids, vitamins, minerals, water and roughage in the **correct proportions** to supply the body with enough **energy** for daily activities and the correct materials for **growth** and **development**, and to keep the body in a **healthy state**.

Carbohydrates, proteins and lipids

Carbohydrates, **proteins** and **lipids** are **organic compounds** that are required in relatively large amounts in a balanced diet, i.e. they are **macronutrients**.

Table 7.2 *Sources and functions of carbohydrates, proteins and lipids*

Class	Sources	Functions
Carbohydrates	Sweet foods, e.g. fruits, cakes, jams. Starchy foods, e.g. yams, potatoes, rice, pasta, bread.	• To provide **energy** (17 kJ g^{-1}): energy is easily released when respired. • For **storage**: glycogen granules are stored in many cells.
Proteins	Fish, lean meat, milk, cheese, eggs, peas, beans, nuts.	• To make **new cells** for growth and to repair damaged tissues. • To make **enzymes** which catalyse reactions in the body. • To make **hormones** which control various processes in the body. • To make **antibodies** to fight disease. • To provide **energy** (17 kJ g^{-1}): used only when stored carbohydrates and lipids have been used up.
Lipids	Butter, vegetable oils, margarine, nuts, fatty meats.	• To make **cell membranes** of newly formed cells. • To provide **energy** (39 kJ g^{-1}): used after carbohydrates because their metabolism is more complex and takes longer. • For **storage**: fat is stored under the skin and around organs. • For **insulation**: fat under the skin acts as an insulator.

Vitamins

Vitamins are **organic compounds** that are required in small amounts for healthy growth and development, i.e. they are **micronutrients**.

Table 7.3 *Some important vitamins required by the human body*

Vitamin	Sources	Functions	Results of deficiency
A	Liver, cod liver oil, yellow and orange vegetables and fruits, e.g. carrots and pumpkin; green leafy vegetables, e.g. spinach.	• Helps to keep the skin, cornea and mucous membranes healthy. • Helps vision in dim light (night vision). • Strengthens the immune system.	• Dry, unhealthy skin and cornea. • Increased susceptibility to infection. • Reduced vision at night or complete night blindness. • **Xerophthalmia**: eyes fail to produce tears leading to a dry, damaged cornea and sometimes blindness.
B_1	Whole-grain cereals and bread, brown rice, peas, beans, nuts, yeast extract, lean pork.	• Aids in respiration to produce energy. • Important for the proper functioning of the nervous system.	• **Beri-beri**: weakness and pain in the limb muscles, difficulty walking, nervous system disorders, paralysis.
B_3	Fish, lean meats, whole-grain cereals, yeast extract.	• Aids in respiration to produce energy.	• **Pellagra**: skin, digestive system and nervous system disorders resulting in dermatitis, diarrhoea and dementia.
C	West Indian cherries, citrus fruits, raw green vegetables.	• Keeps tissues healthy, especially the skin and connective tissue. • Strengthens the immune system.	• **Scurvy**: swollen and bleeding gums, loose teeth or loss of teeth, red-blue spots on the skin, muscle and joint pain, wounds do not heal. • Increased susceptibility to infection.
D	Oily fish, eggs, cod liver oil. Made in the body by the action of sunlight on the skin.	• Promotes the absorption of calcium and phosphorus in the ileum. • Helps build and maintain strong bones and teeth. • Strengthens the immune system.	• **Rickets** in children: soft, weak, painful, deformed bones, especially limb bones, bow legs. • **Osteomalacia** in adults: soft, weak, painful bones that fracture easily, weakness of limb muscles. • Poor teeth.

Vitamins A and D are **fat soluble**
Vitamins B and C are **water soluble**

Results of vitamin surplus

Vitamins, especially vitamins A and D, can become **harmful** to the body when consumed in excess. This can occur when taking supplements, especially in children.

- A surplus of **vitamin A** can cause liver damage, jaundice, itchy skin, cracked fingernails, blurry vision, nausea, headaches and fatigue.
- A surplus of **vitamin D** can cause high levels of calcium in the blood, excessive thirst and urination, loss of appetite, nausea, vomiting, calcification of soft tissues, e.g. kidneys, lungs and inside blood vessels, and the development of kidney stones.

Minerals

Minerals are **inorganic substances** that are required in small amounts for healthy growth and development, i.e. they are **micronutrients**.

Table 7.4 *Some important minerals required by the human body*

Element	Sources	Functions	Results of deficiency
Calcium (Ca)	Dairy products, e.g. milk, cheese and yoghurt; green vegetables, e.g. broccoli.	• To build and maintain healthy bones and teeth. • Helps blood to clot at cuts.	• **Rickets** in children. • **Osteoporosis** in adults: brittle, fragile bones. • Weak, brittle nails. • Tooth decay.
Phosphorus (P)	Protein-rich foods, e.g. milk, cheese, meat, poultry, fish, nuts.	• To build and maintain healthy bones and teeth. • To make ATP, an energy-rich compound (see page 64).	• Weak bones and teeth. • Tiredness, lack of energy.
Iron (Fe)	Red meat, liver, eggs, beans, nuts, dark green leafy vegetables.	• To make haemoglobin, the red pigment in red blood cells.	• **Anaemia**: reduced numbers of red blood cells in the blood, pale complexion, tiredness, lack of energy.
Iodine (I)	Sea foods, e.g. fish, shellfish and seaweed, milk, eggs.	• To make the hormone thyroxine.	• **Cretinism** in children: retarded physical and mental development. • **Goitre** in adults: swollen thyroid gland in the neck. • Reduced metabolic rate leading to fatigue in adults.
Sodium (Na), **potassium** (K)	Sodium: table salt, cheese, cured meats. Potassium: fruits, vegetables.	• Needed for the transmission of nerve impulses and muscle contraction. • Help maintain the correct concentration of body fluids.	• Muscular cramps.
Fluorine (F)	Fluoridated tap water, fluoride toothpaste.	• Strengthens tooth enamel making it more resistant to decay.	• Teeth decay more rapidly than normal.

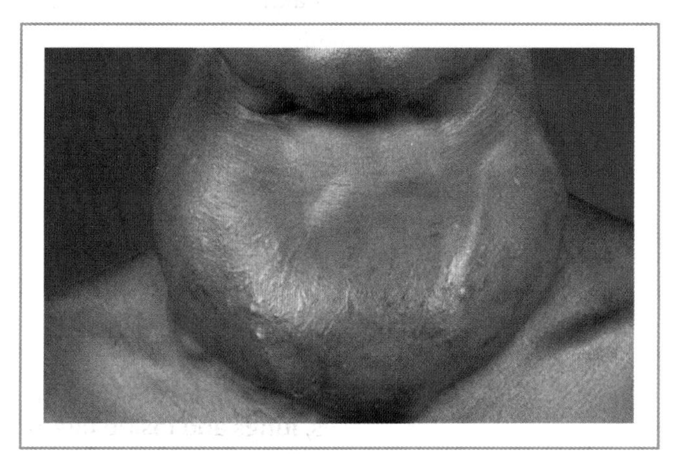

Goitre

Results of mineral surplus

Minerals can become **harmful** to the body when consumed in excess. This can occur when taking supplements.

- A surplus of **calcium** can cause calcification of soft tissues, especially the kidneys and inside blood vessels, and the development of kidney stones.
- A surplus of **sodium** can raise blood pressure resulting in hypertension, cause the body to retain fluid and cause kidney damage.
- A surplus of **iron** can lead to liver damage.

Water

Water is an inorganic compound that is essential in a balanced diet. The human body is about 65% water.

- Water **dissolves** chemicals in cells so that they can **react**.
- Water **dissolves** substances so that they can be **transported** around the body, e.g. products of digestion are dissolved in blood plasma.
- Water **dissolves** waste substances so that they can be **excreted** from the body, e.g. urea.
- Water acts as a **reactant**, e.g. in **hydrolysis** which occurs during digestion.
- Water acts as a **coolant**, removing heat from the body when it **evaporates** from sweat.

Roughage (dietary fibre)

Roughage is food that **cannot be digested**. It consists mainly of the cellulose of plant cell walls, lignin of plant xylem vessels, husks of brown rice and bran of whole-grain cereals. Roughage adds **bulk** to the food, which stimulates **peristalsis** (see page 60) so that food is kept moving through the digestive system. This helps prevent constipation and reduces the risk of colon cancer.

Fruit and cereals are rich in roughage

Energy requirements

The amount of energy required daily from the diet depends on a person's **age**, **occupation** and **gender (sex)**. In general, daily energy requirements:

- **Increase** as **age increases** up to adulthood. They then remain fairly constant up to old age when less energy is required daily.
- **Increase** as **activity increases**, e.g. a manual labourer requires more energy than a person working in an office.
- Are **higher** in **males** than in females of the same age and occupation.
- **Increase** in a female when she is **pregnant** or **breast feeding**.

Malnutrition

Malnutrition is a condition caused by eating a diet in which certain nutrients are either **lacking**, are in **excess** or are in the **wrong proportions**. In general:

- If **too little food** is eaten to meet the body's daily energy requirements, stored glycogen and fat are used in respiration resulting in **weight loss** and **insufficient energy** for daily activities. In extreme cases it can lead to **marasmus**, a condition where the body wastes away.
- If **too much food** is eaten, the excess is converted to **fat** and stored in fat deposits under the skin and around organs. This results in **overweight** and **obesity**, and can lead to **diabetes**, **hypertension** and **heart disease**.
- If certain nutrients are consumed in the **wrong proportions** this can lead to malnutrition, e.g. **kwashiorkor** in children is caused by a deficiency of protein. Consumption of too little or too much of certain vitamins and minerals also leads to nutritional disorders (see pages 52 to 54).

Diet and the treatment and control of disease

Deficiency diseases and certain **physiological diseases** (see page 135) can be treated and controlled by making changes to the diet.

- **Deficiency diseases** such as rickets, scurvy, anaemia and kwashiorkor can be treated by increasing the intake of foods **rich** in the missing nutrient or foods **fortified** with the missing nutrient, or by taking **supplements** containing the missing nutrient.

Scurvy is treated by taking vitamin C supplements or eating foods rich in vitamin C, such as oranges

- **Diabetes** can be controlled by eating a healthy, balanced diet that is low in sugar and saturated fats and high in dietary fibre supplied by fresh fruits, vegetables and whole grains. In particular, people with diabetes should consume foods containing polysaccharides rather than simple sugars, and fish and lean meat rather than fatty meats.
- **Hypertension** (high blood pressure) can be controlled by eating a balanced diet that is low in saturated fat, cholesterol and salt, and high in dietary fibre, potassium, calcium and magnesium. The diet should contain plenty of fresh fruits, vegetables and whole grains together with low-fat dairy products, fish and lean meat. Persons suffering from hypertension should also stop smoking, reduce obesity and reduce alcohol consumption.

Vegetarianism

Vegetarianism is the practice of not eating the flesh of any animal, e.g. meat, fish, poultry. Strict vegetarians do not consume any foods of animal origin, e.g. milk, eggs, cheese. A vegetarian diet needs to be more **carefully planned** than a non-vegetarian diet to ensure it is balanced. Once planned properly, a vegetarian diet has advantages:

- The diet is low in **saturated fats** and **cholesterol**, therefore, vegetarians are less prone to obesity, heart disease, hypertension, diabetes and gall stones.
- The diet is high in **dietary fibre**, therefore, vegetarians are less likely to suffer from constipation, colon cancer and certain other types of cancer.

Revision questions

9 It is essential that humans consume a balanced diet daily. What is a balanced diet and why is it needed?

10 Why is it important for a person's daily diet to contain sufficient protein?

11 Construct a table to give ONE source, the functions and the effects of the deficiency of the following micronutrients: vitamin C, vitamin D, iron and iodine.

12 What is the importance of roughage in the daily diet?

13 What factors affect a person's daily energy requirements?

14 Suggest THREE causes of malnutrition.

15 What dietary advice would you give to a person suffering from:
a diabetes **b** hypertension **c** a deficiency disease?

16 Suggest TWO benefits of a vegetarian diet.

Digestion in humans

Digestion breaks down food into a form that is useful for body activities, i.e. simple, soluble molecules. It occurs in the **alimentary canal** which is a tube 8 to 9 metres long with muscular walls running from the **mouth** to the **anus**. Different regions of the canal are adapted to perform different functions (see Figure 7.6, page 59).

Digestion begins with the **mechanical** breakdown of pieces of food and this is followed by the **chemical** breakdown of food molecules.

Mechanical digestion

Mechanical digestion involves breaking up **large pieces** of food into **smaller pieces**. This is important because:

- It gives the pieces of food a **larger surface area** for digestive enzymes to act on, making chemical digestion quicker and easier.
- It makes food easier to **swallow**.

Mechanical digestion begins in the **mouth** where it is carried out mainly by the **teeth**. Humans have **four** different types of teeth.

Table 7.5 *The different types of teeth in humans*

Type	Position	Shape	Functions
Incisor	At the front of the jaw.	Chisel-shaped with sharp, thin edges. 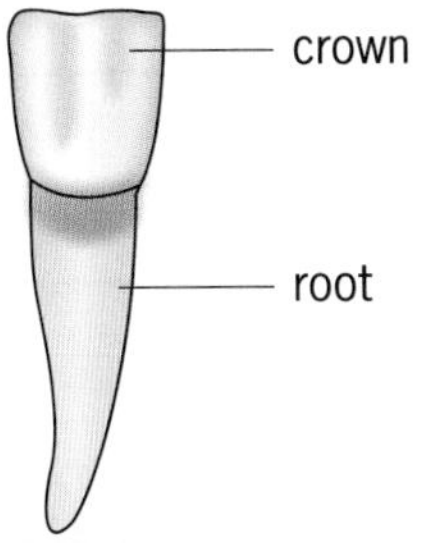 crown root	To cut food. To bite off pieces of food.
Canine (eye tooth)	Next to the incisors.	Cone-shaped and pointed. 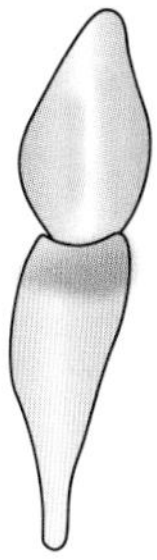	To grip food. To tear off pieces of food.
Premolar	At the side of the jaw next to the canines.	Have a fairly broad surface with two pointed cusps. 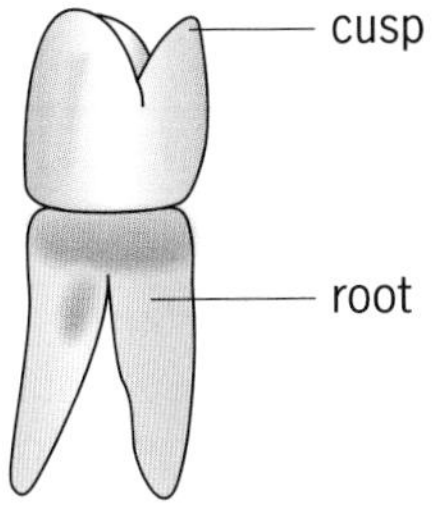 cusp root	To crush and grind food.
Molar	At the back of the jaw next to the premolars.	Have a broad surface with 4 or 5 pointed cusps. 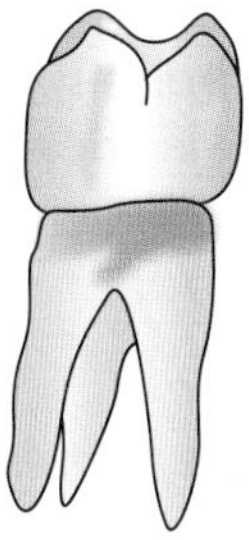	To crush and grind food.

A tooth is divided into **two** parts; the **crown** which is the part above the jaw and the **root** which is the part embedded in the jawbone. The internal structure of all teeth is similar.

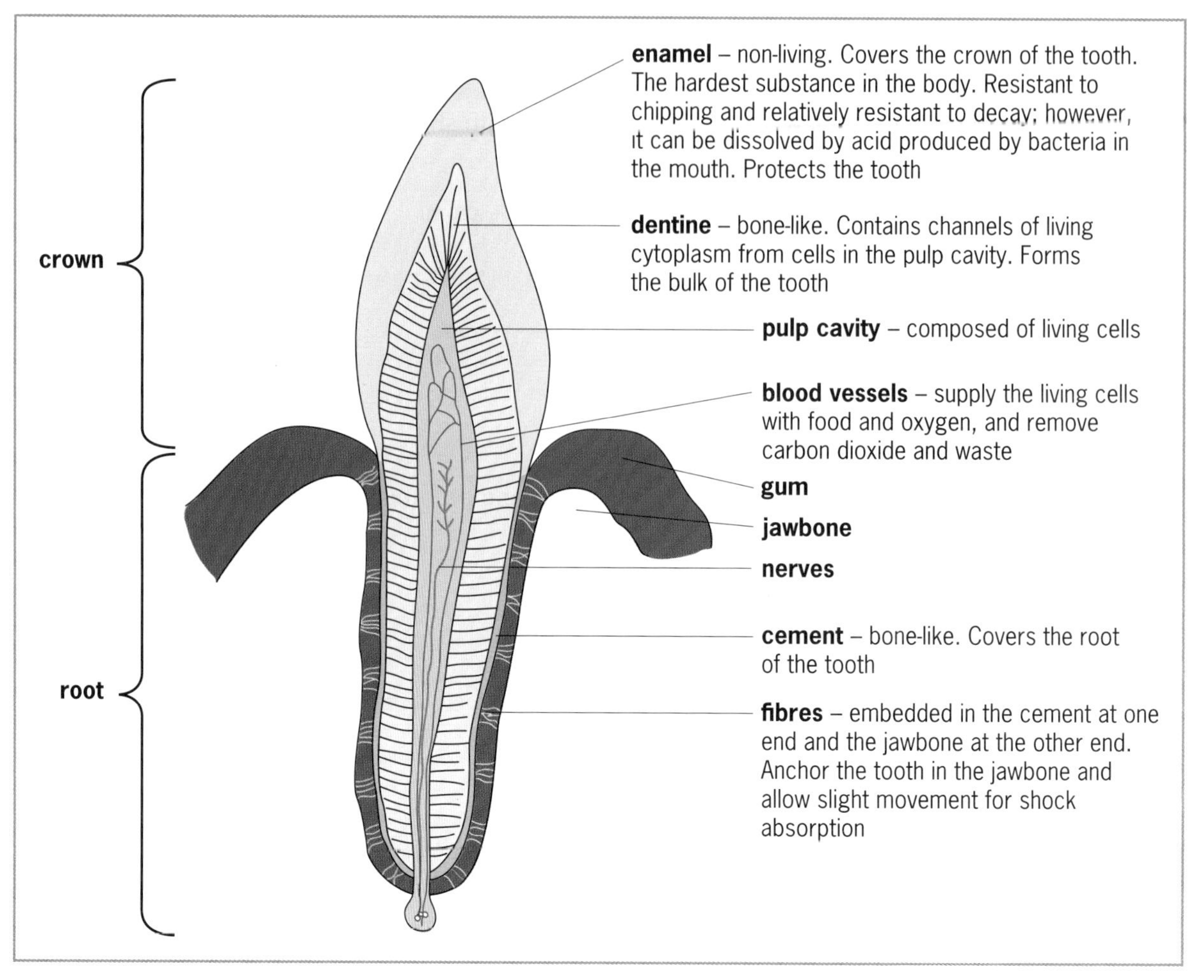

Figure 7.5 *Structure and function of the parts of a canine tooth as seen in longitudinal section*

When chewing is complete, the food is rolled into a ball called a **bolus**. The bolus is pushed to the back of the mouth by the tongue and swallowed. It enters the **stomach** where contractions of the muscles in the stomach walls churn the food which continues the process of mechanical digestion.

Chemical digestion

Chemical digestion involves breaking down large, usually insoluble, food molecules into small, soluble food molecules by **hydrolysis**. During hydrolysis, the bonds within the large food molecules are broken down by the addition of **water** molecules. Hydrolysis is catalysed by **digestive enzymes**. There are **three** categories of digestive enzymes and several different enzymes may belong to each category (see Tables 7.6 and 7.7, below).

Table 7.6 *Categories of digestive enzymes*

Category of digestive enzyme	Food molecules hydrolysed	Products of hydrolysis
Carbohydrases	Polysaccharides and disaccharides	Monosaccharides
Proteases	Proteins	Amino acids
Lipases	Lipids	Fatty acids and glycerol

Chemical digestion begins in the **mouth**, continues in the **stomach** and is completed in the **small intestine**.

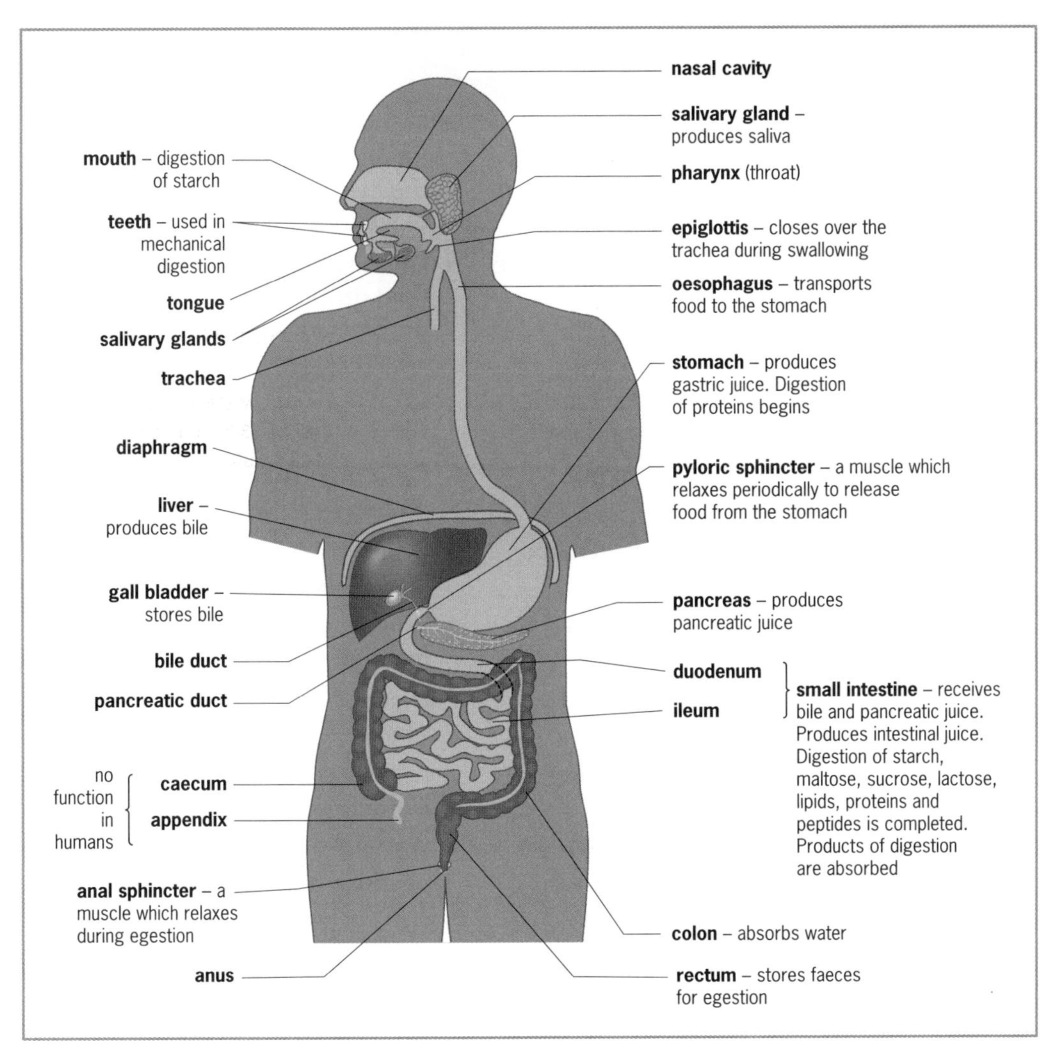

Figure 7.6 *Functions of the parts of the human alimentary canal summarised*

Table 7.7 *Chemical digestion in humans summarised*

Organ	Digestive juice	Source	Main components	Functions of the components
Mouth	Saliva	Salivary glands.	• Water and mucus	• Moisten and lubricate the food allowing tasting and easy swallowing.
			• **Salivary amylase***	• Begins to digest: **starch ⟶ maltose**
Stomach	Gastric juice	Cells in the stomach wall.	• Hydrochloric acid	• Maintains an optimum pH of 1–2 for pepsin and rennin, and kills bacteria.
			• **Rennin***	• Produced in infants to clot soluble protein in milk so the protein is retained in the stomach.
			• **Pepsin***	• Begins to digest: **protein ⟶ peptides** (shorter chains of amino acids)

Organ	Digestive juice	Source	Main components	Functions of the components
Small intestine (duodenum and ileum)	Bile	Cells in the liver. It is stored in the gall bladder and enters the duodenum via the bile duct.	• Bile pigments, e.g. bilirubin • Organic bile salts	• Excretory products from the breakdown of haemoglobin in the liver. Have no function in digestion. • **Emulsify lipids**, i.e. break large lipid droplets into smaller droplets increasing their surface area for digestion.
	Pancreatic juice	Cells in the pancreas. It enters the duodenum via the pancreatic duct.	• **Pancreatic amylase*** • **Trypsin*** • **Pancreatic lipase***	• Continues to digest: **starch** ⟶ **maltose** • Continues to digest: **protein** ⟶ **peptides** • Digests: **lipids** ⟶ **fatty acids** and **glycerol**
	Intestinal juice	Cells in the walls of the small intestine.	• **Maltase*** • **Sucrase*** • **Lactase*** • **Peptidase* (erepsin)**	• Digests: **maltose** ⟶ **glucose** • Digests: **sucrose** ⟶ **glucose** and **fructose** • Digests: **lactose** ⟶ **glucose** and **galactose** • Digests: **peptides** ⟶ **amino acids**

* digestive enzymes

Movement of food through the alimentary canal

Food is moved through the oesophagus and the rest of the alimentary canal by a process known as **peristalsis**.

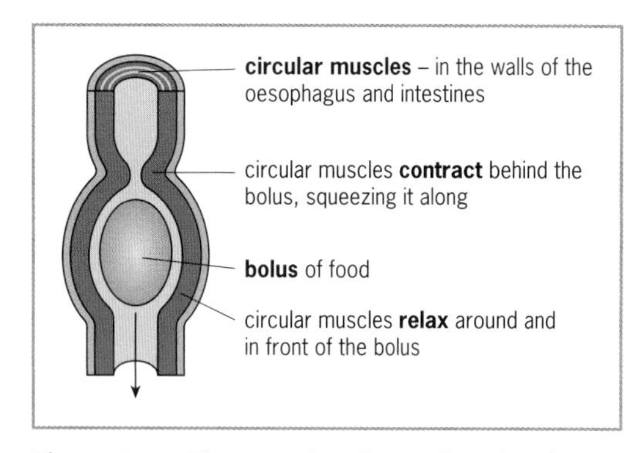

Figure 7.7 *The mechanism of peristalsis*

Absorption

Absorption of the products of digestion into body fluids occurs in the **small intestine** and **colon**.

Absorption in the small intestine

The products of digestion are **absorbed** through the lining of the small intestine, mainly the ileum, and into the blood vessels and lymphatic vessels in its walls. Substances absorbed include monosaccharides, amino acids, fatty acids, glycerol, vitamins, minerals and water.

Absorption occurs by both **diffusion** and **active transport** (see pages 37 and 39). The ileum is **adapted** to absorb food as **efficiently** as possible:

- It is **very long**, about 5 m in an adult. This provides a **large surface area** for rapid absorption.
- Its inner surface has thousands of finger-like projections known as **villi** (singular villus). These greatly increase the **surface area** for absorption.
- Each villus has a network of **blood capillaries** and a **lacteal** inside. These provide means of rapidly **transporting** the products of digestion away from the ileum.
- The **wall** of each villus, known as the **epithelium**, is only **one cell thick**. Digested food can pass rapidly through this epithelium into the capillaries and lacteal.
- The epithelial cells have tiny projections called **microvilli**. These further increase the **surface area** for absorption.

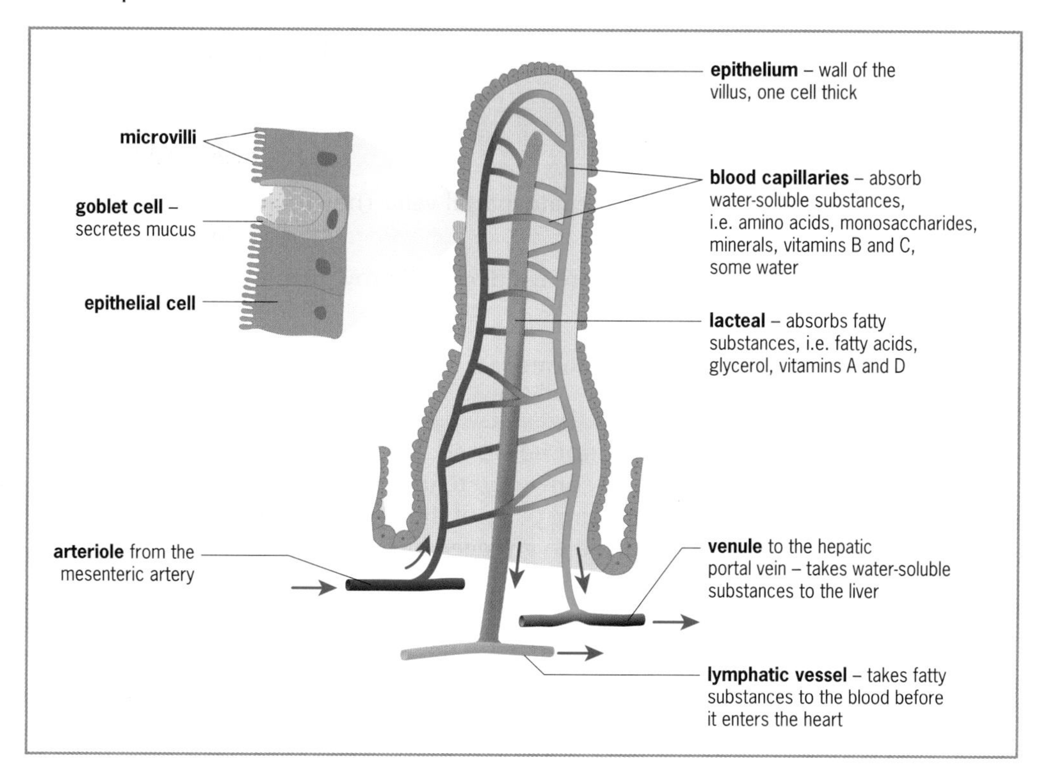

Figure 7.8 *Structure of a villus showing absorption*

Absorption in the colon

Any food that is not digested in the small intestine passes into the **colon** where **water** and **mineral salts** are absorbed from it. As this undigested waste moves along the colon to the rectum it becomes progressively more solid as the water is absorbed.

Egestion

The almost solid material entering the rectum is called **faeces** and consists of undigested dietary fibre, dead bacteria and intestinal cells, mucus and bile pigments. Faeces is stored in the rectum and **egested** at intervals through the **anus** when the **anal sphincter** relaxes.

Assimilation

The body **uses** the products of digestion in a variety of ways.

Monosaccharides

Monosaccharides are taken by the blood to the liver in the **hepatic portal vein** and the liver converts any non-glucose monosaccharides to **glucose**. The glucose then enters the general circulation where:

- It is used by all body cells in **respiration** to produce **energy**.
- Excess is condensed to **glycogen** by cells in the **liver** and **muscles**. These cells then store the glycogen, or
- Excess is converted to **fat** by cells in the **liver** and **adipose tissue** found under the skin and around organs. Fat made in adipose tissue is stored, and fat made in the liver is transported by the blood to adipose tissue and stored.

Amino acids

Amino acids are taken by the blood to the liver in the **hepatic portal vein**. They then enter the general circulation where:

- They are used by body cells to make **proteins** which are used for cell growth and repair.
- They are used by body cells to make **enzymes**.
- They are used by cells of endocrine glands to make **hormones**.
- They are used to make **antibodies**.
- Excess are **deaminated** by the **liver** because they cannot be stored. The nitrogen-containing amine groups (NH_2) are removed from the molecules and converted to **urea** ($CO(NH_2)_2$). The urea enters the blood and is **excreted** by the **kidneys**. The remaining parts of the molecules are converted to **glucose** which is used in respiration, or are converted to **glycogen** or **fat** and stored.

Fatty acids and glycerol

Fatty acids and **glycerol** are carried by the **lymph** to the general circulation where:

- They are used to make **cell membranes** of newly forming cells.
- They are used by body cells in **respiration** under some circumstances.
- Excess are converted to **fat** and stored in adipose tissue under the skin and around organs.

Control of blood glucose levels

Two **hormones**, secreted directly into the blood by the **pancreas**, are responsible for keeping blood glucose levels constant. These are **insulin** and **glucagon**. Regulation of blood glucose levels is one aspect of **homeostasis**, i.e. the maintenance of a constant internal environment.

- If the blood glucose level **rises**, e.g. after a meal rich in carbohydrates, the pancreas secretes **insulin** which stimulates body cells to absorb glucose for **respiration** and the liver cells to convert excess glucose to **glycogen**.
- If the blood glucose level **falls**, e.g. between meals, or during exercise or sleep, the pancreas secretes **glucagon** which stimulates the liver cells to convert glycogen to **glucose** which enters the blood.

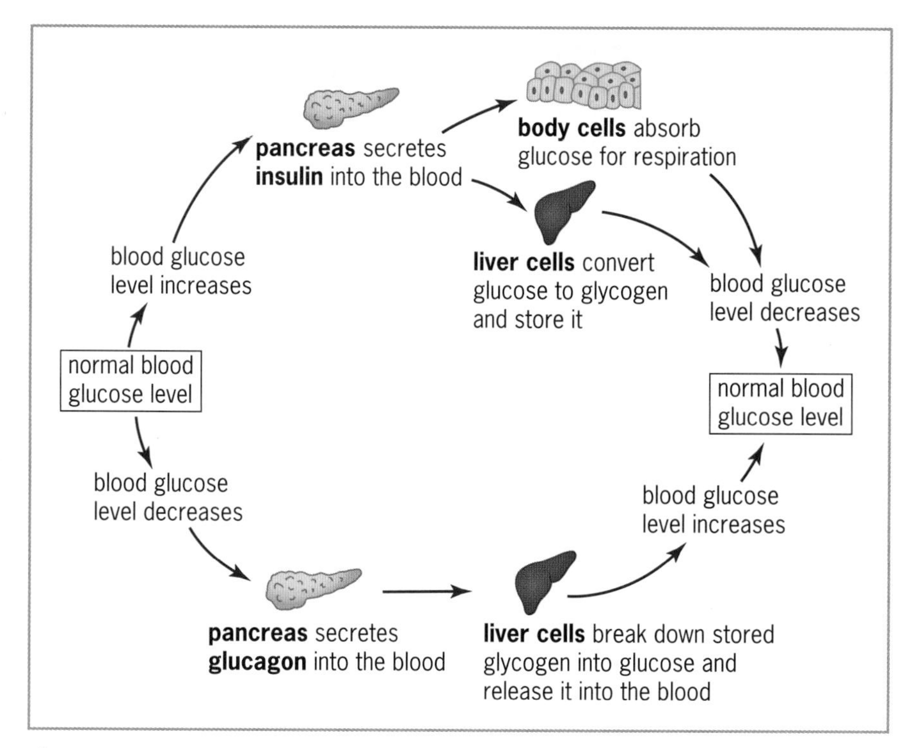

Figure 7.9 *Control of blood glucose levels*

Revision questions

17 What happens during digestion?

18 By means of a fully labelled and annotated diagram only, describe the internal structure of a canine tooth.

19 State the function of any THREE of the structures you have labelled in question **18**.

20 Why are teeth important in the digestive process?

21 For lunch you consume a ham sandwich that contains lettuce and tomatoes. Describe how this sandwich is digested as it passes through your digestive system.

22 The products of the digestion of your ham sandwich are absorbed in your small intestine. Explain FOUR ways in which your small intestine is adapted to efficiently perform its function of absorption.

23 What happens to any excess amino acids in the diet?

24 After a meal rich in carbohydrates, the blood glucose level rises. How does the body function to return the blood glucose level to normal?

8 Respiration and gaseous exchange

All living organisms need **energy** to carry out life processes in order to survive. They obtain this energy from food.

Respiration

Respiration *is the process by which energy is released from food by all living cells.*

Respiration is catalysed by **enzymes** and occurs slowly in a large number of stages. Energy released at each stage is used to build **energy carrying** molecules called **adenosine triphosphate** or **ATP**. An ATP molecule is formed by combining some of the **energy** released with an **adenosine diphosphate** or **ADP** molecule and a **phosphate** group present in the cell. The energy can then be re-released wherever it is needed in the cell by the reverse reaction:

$$\text{ADP} + \text{phosphate} + \text{energy} \rightleftharpoons \text{ATP}$$

ATP is known as the '**energy currency**' of cells. Cells earn ATP as a result of energy-producing reactions and spend it on reactions requiring energy. Storage and transport of energy in ATP has the following advantages:

- The energy can be released **rapidly**.
- Only a **small amount** of energy is released when each ATP molecule is broken down. This **prevents waste** since it allows the release of exactly the **right amount** of energy when needed.
- Energy can be released exactly **where** it is needed in the cell.

Energy released by ATP is **used** by cells:

- To manufacture complex, biologically important molecules, e.g. proteins, DNA.
- For cell growth and repair.
- For cell division.
- In active transport to move molecules and ions in and out of the cells through their membranes.
- For special functions in specialised cells, e.g. contraction of muscle cells and the transmission of impulses in nerve cells.

Not all the energy released in respiration is converted to ATP. Some is released as **heat** which helps maintain the body temperature at 37 °C.

There are **two** types of respiration:

- **aerobic respiration**
- **anaerobic respiration**.

Aerobic respiration

Aerobic respiration occurs in most cells. It **uses oxygen** and takes place in the **mitochondria**. It always produces **carbon dioxide**, **water** and about **38 ATP molecules** per molecule of glucose.

$$\text{glucose} + \text{oxygen} \xrightarrow[\text{mitochondria}]{\text{enzymes in}} \text{carbon dioxide} + \text{water} + \text{energy}$$

or

$$C_6H_{12}O_6 + 6O_2 \xrightarrow[\text{mitochondria}]{\text{enzymes in}} 6CO_2 + 6H_2O + \text{energy}$$

Anaerobic respiration

Anaerobic respiration occurs in some cells. It takes place **without oxygen** in the **cytoplasm** of the cells and produces considerably less energy per molecule of glucose than aerobic respiration.

- **Anaerobic respiration in yeast cells**

  Yeast cells carry out anaerobic respiration known as **fermentation**. It produces **ethanol, carbon dioxide** and **2 ATP molecules** per molecule of glucose.

$$\text{glucose} \xrightarrow[\text{cytoplasm}]{\text{enzymes in}} \text{ethanol} + \text{carbon dioxide} + \text{energy}$$

or

$$C_6H_{12}O_6 \xrightarrow[\text{cytoplasm}]{\text{enzymes in}} 2C_2H_5OH + 2CO_2 + \text{energy}$$

- **Anaerobic respiration in muscle cells**

  During **strenuous exercise** when the oxygen supply to muscle cells becomes too low for the demands of aerobic respiration the cells begin to respire **anaerobically**. This produces **lactic acid** and **2 ATP molecules** per molecule of glucose.

$$\text{glucose} \xrightarrow[\text{cytoplasm}]{\text{enzymes in}} \text{lactic acid} + \text{energy}$$

or

$$C_6H_{12}O_6 \xrightarrow[\text{cytoplasm}]{\text{enzymes in}} C_3H_6O_3 + \text{energy}$$

  Lactic acid builds up in the muscle cells and begins to harm them causing fatigue and eventually collapse as they stop contracting. The muscle cells are said to have built up an **oxygen debt**. This debt must be **repaid** directly after exercise by resting and breathing deeply so that the lactic acid can be respired **aerobically**.

- **Anaerobic respiration in certain bacteria**

  Certain **bacteria** obtain energy by breaking down **organic waste**, e.g. manure and garden waste, under anaerobic conditions to produce **biogas**. Biogas is a mixture of approximately **60% methane (CH_4)**, **40% carbon dioxide** and traces of other contaminant gases such as hydrogen sulfide (H_2S).

Table 8.1 *Aerobic and anaerobic respiration compared*

Aerobic respiration	Anaerobic respiration
Uses oxygen.	Does **not** use oxygen.
Occurs in the **mitochondria** of cells.	Occurs in the **cytoplasm** of cells.
Releases **large amounts** of energy: about 38 ATP molecules are produced per glucose molecule.	Releases **small amounts** of energy: 2 ATP molecules are produced per glucose molecule.
The products are always **inorganic**, i.e. carbon dioxide and water.	The products are variable, at least one is always **organic**, e.g. ethanol, lactic acid or methane.
Glucose is **completely** broken down and the products contain no energy.	Glucose is only **partially** broken down. The organic products still contain energy.

Revision questions

1. What is respiration?
2. What is ATP and what are the advantages of cells producing ATP?
3. Distinguish between aerobic respiration and anaerobic respiration and write a balanced chemical equation to summarise the process of aerobic respiration.
4. Why does anaerobic respiration release less energy per molecule of glucose than aerobic respiration?
5. An athlete carried out strenuous exercise for an extended period of time and eventually collapsed and found he had to rest before resuming any exercise. Explain why he collapsed and why he had to rest before he could exercise again.
6. What is biogas and how is it produced?

Gaseous exchange and breathing

Gaseous exchange is the process by which oxygen **diffuses** into an organism, and carbon dioxide diffuses out of an organism, through a **gaseous exchange surface (respiratory surface)**.

Breathing refers to the **movements** in animals that bring oxygen to a gaseous exchange surface (see below) and remove carbon dioxide from the surface. Breathing must not be confused with respiration.

Breathing and gaseous exchange are essential to organisms that respire **aerobically**:

- To ensure they have a **continual supply** of **oxygen** to meet the demands of aerobic respiration.
- To ensure that the **carbon dioxide** produced in respiration is **continually removed** so that it does not build up and poison cells.

Gaseous exchange surfaces

In many animals the gaseous exchange surface forms part of the **respiratory system**. These surfaces have several **adaptations** that make the exchange of gases through them as **efficient** as possible:

- They have a **large surface area** so that large quantities of gases can be exchanged.
- They are very **thin** so that gases can diffuse through them rapidly.
- They are **moist** so that gases can dissolve before they diffuse through the surface.
- They have a **rich blood supply** (if the organism has blood) to quickly transport gases between the surface and the body cells.

Breathing and gaseous exchange in humans

Gaseous exchange occurs in the **lungs** of humans. Humans have **two** lungs composed of thousands of air passages called **bronchioles** and millions of swollen air sacs called **alveoli**. Each lung is surrounded by two **pleural membranes** which have **pleural fluid** between, and a single **bronchus** leads into each from the **trachea**. Each lung receives blood from the heart via a **pulmonary artery** and blood is carried back to the heart via a **pulmonary vein**.

The two lungs are surrounded by the **ribs** which form the **chest cavity** or **thorax**. The ribs have **intercostal muscles** between, and a dome-shaped sheet of muscle, the **diaphragm**, stretches across the floor of the thorax. Movements of the ribs and diaphragm, brought about by the muscles contracting and relaxing, cause air to move in and out of the lungs.

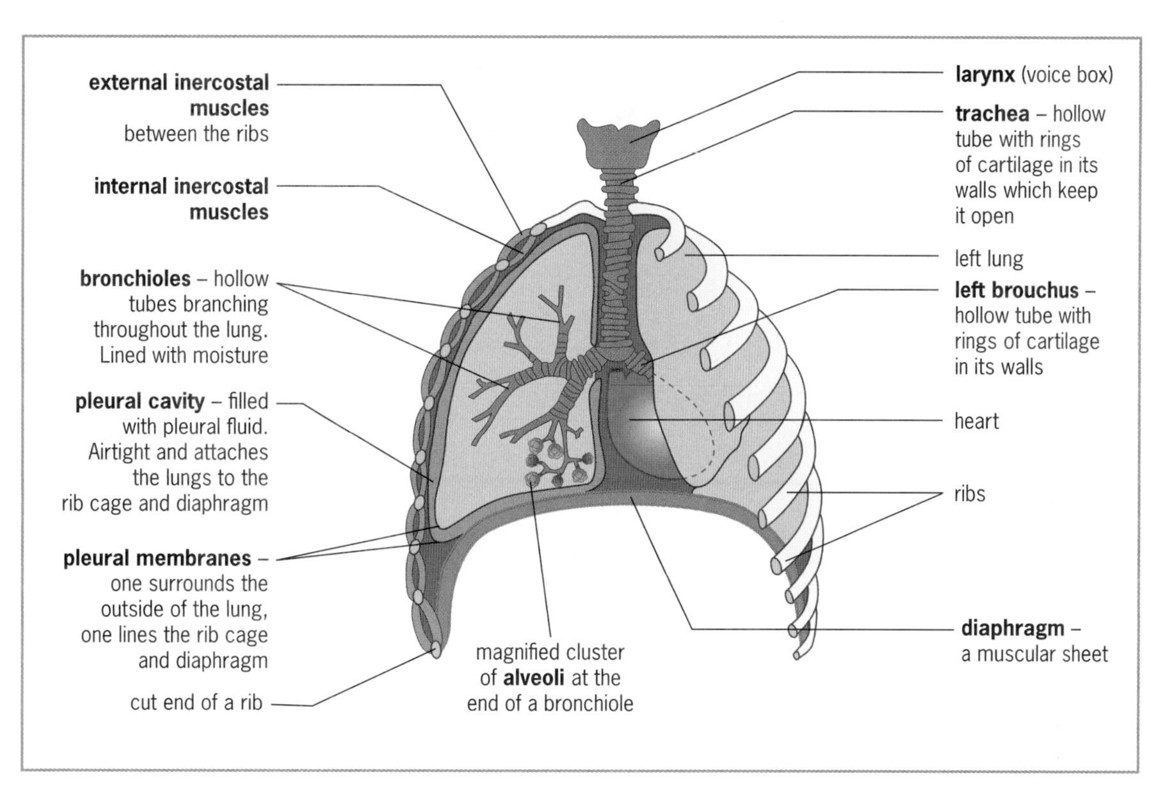

Figure 8.1 *Structure of the human respiratory system*

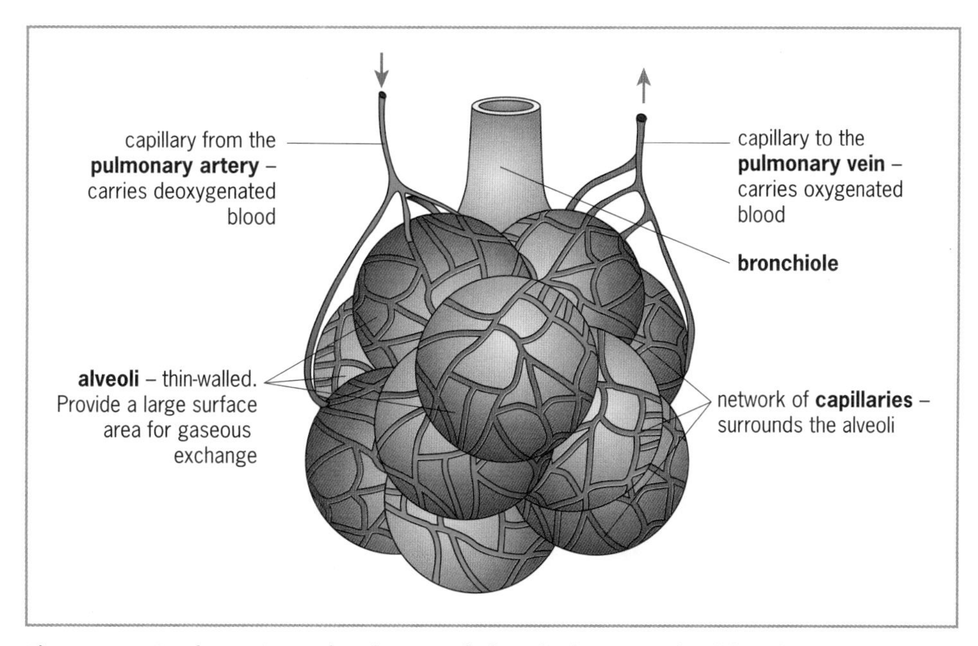

Figure 8.2 *Surface view of a cluster of alveoli showing the blood supply*

The **walls** of the **alveoli** form the gaseous exchange surface. Each **alveolus**:

- Has a wall that is one cell thick.
- Is lined with moisture.
- Is surrounded by a network of capillaries.

Table 8.2 *The mechanism of breathing in humans*

	Features	Inhalation (inspiration)	Exhalation (expiration)
1	External intercostal muscles	Contract	Relax
	Internal intercostal muscles	Relax	Contract
	Ribs and sternum	Move upwards and outwards	Move downwards and inwards
2	Diaphragm muscles	Contract	Relax
	Diaphragm	Moves downwards or flattens	Domes upwards
3	Volume inside thorax and lungs	Increases	Decreases
	Pressure inside thorax and lungs	Decreases	Increases
4	Movement of air	Air is drawn into the lungs due to the decrease in pressure	Air is pushed out of the lungs due to the increase in pressure

As the air is drawn in during inhalation, it is **warmed** in the nasal passages, and **cleaned** and **moistened** by mucus lining the nasal passages and trachea. The mucus is moved to the throat by **cilia** (microscopic hairs) lining the nasal passages and trachea, and is swallowed. The air passes through the bronchi and bronchioles and enters the alveoli where **gaseous exchange** occurs between the air and the blood in the capillaries.

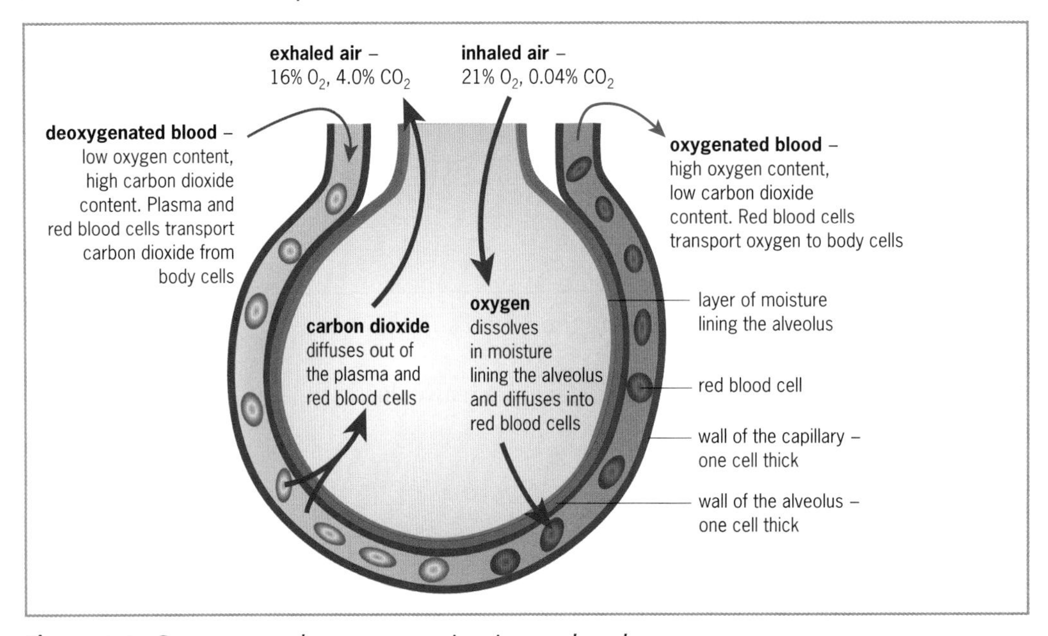

Figure 8.3 *Gaseous exchange occurring in an alveolus*

Table 8.3 *The composition of inhaled and exhaled air* compared

	Inhaled air	Exhaled air	Reason for the differences
Oxygen (O_2)	21%	16%	Oxygen diffuses into the blood from the inhaled air, is transported to the body cells and used in respiration.
Carbon dioxide (CO_2)	0.04%	4%	Carbon dioxide, produced by the body cells during respiration, diffuses out of the blood into the air and is exhaled.
Nitrogen (N_2)	78%	78%	Nitrogen is not used by body cells.
Water vapour (H_2O)	Variable	Saturated	The gaseous exchange surface is moist and some of the moisture evaporates into the air being exhaled.
Temperature	Variable	Body temperature	The air is warmed as it moves through the respiratory system.

The gaseous exchange surface in a fish

Gaseous exchange occurs in the **gills** of a fish. A bony fish has four gills at each side of its head. Each gill has two rows of long, thin, finger-like projections called **gill lamellae**.

The **walls** of the **gill lamellae** form the gaseous exchange surface. Each **lamella**:

- Has a wall that is one cell thick.
- Is moist since it has water passing over it as the fish breathes.
- Has a dense network of capillaries down the centre.

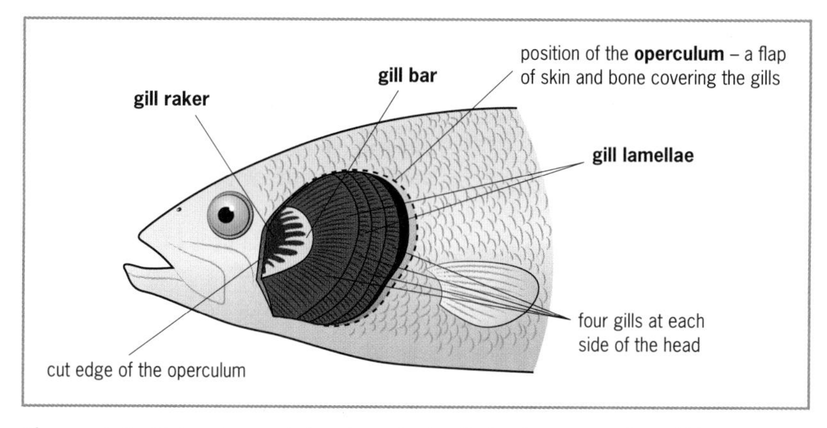

Figure 8.4 *Side view of the head of a fish showing the gills*

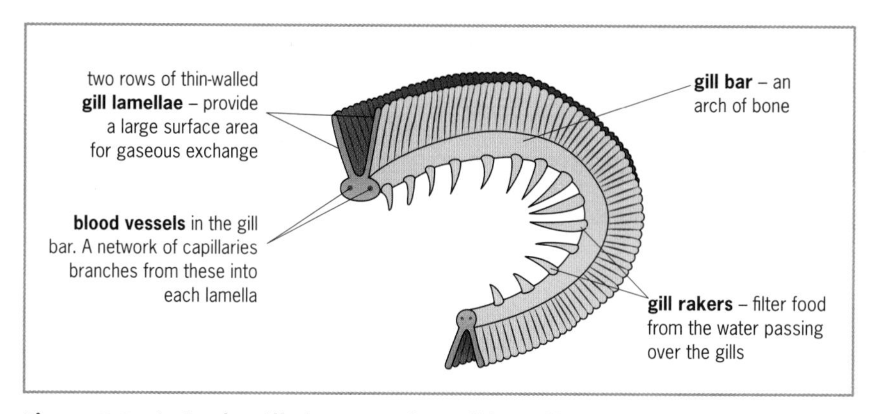

Figure 8.5 *A single gill showing the gill lamellae*

Gaseous exchange in flowering plants

Gaseous exchange occurs in the **leaves**, **stems** and **roots** of plants by **direct diffusion** between the intercellular **air spaces** and all the **cells** in these organs that are in contact with the air spaces. Gases diffuse between the **atmosphere** and the air spaces through the **stomata** of leaves and the **lenticels** of bark covered stems and roots. Lenticels are small areas of loosely packed cells.

The **walls** and **membranes** of all the **cells** inside the leaves, stems and roots of plants form the gaseous exchange surface.

The direction of movement of gases depends on whether the organ of the plant is also carrying out photosynthesis.

Gaseous exchange in photosynthesising organs

Movement of gases into and out of photosynthesising organs, mainly **leaves**, depends on the time of the day. In these organs, the rate of respiration remains almost constant throughout the day and night, but the rate of photosynthesis changes.

- During the **night**, only **respiration** occurs. **Oxygen** diffuses **in** and **carbon dioxide** diffuses **out**.
- As **dawn** approaches photosynthesis begins and its rate gradually increases. The **compensation point** is reached when the rate of photosynthesis **equals** the rate of respiration. At this point, there is **no net movement** of gases in or out of leaves.
- During the **day**, the rate of **photosynthesis** is greater than the rate of respiration. **Carbon dioxide** diffuses **in** and **oxygen** diffuses **out**.
- At about **dusk** a second compensation point occurs as the rates of photosynthesis and respiration become **equal** once more.

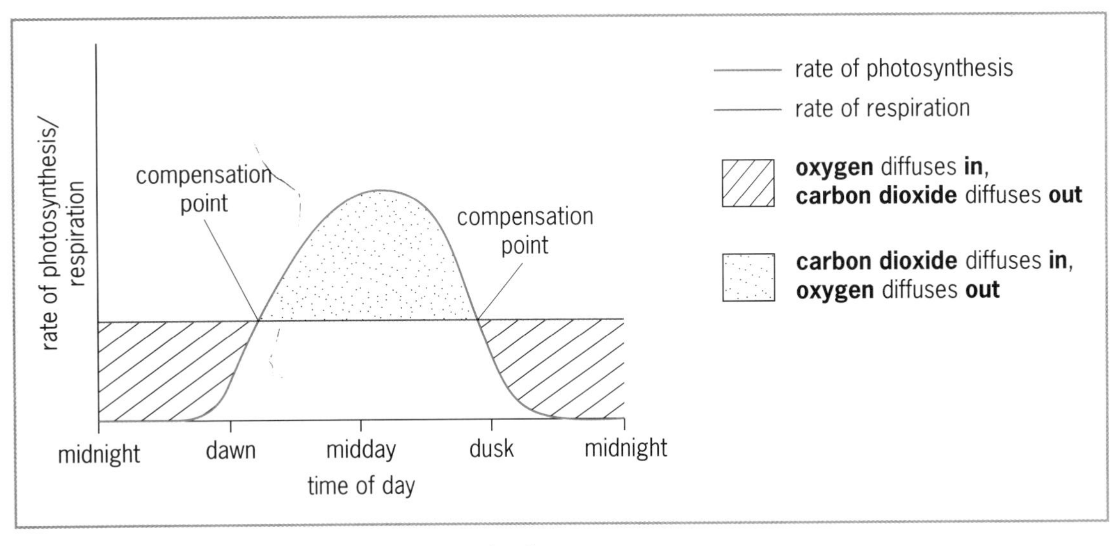

Figure 8.6 *Gaseous exchange in a green leaf*

Gaseous exchange in non-photosynthesising organs

Only **respiration** occurs in non-photosynthesising organs, e.g. stems and roots. **Oxygen** diffuses **in** and **carbon dioxide** diffuses **out** at all times.

Table 8.4 *A summary of the gaseous exchange surfaces of different organisms*

Adaptation of the gaseous exchange surface	Organism		
	Human	**Fish**	**Flowering plant**
Large surface area	Each alveolus has a pocket shape and a human has two lungs, each with over 350 million alveoli giving a total surface area of about 90 m^2.	Each lamella is long and thin and a fish has eight gills, each with a large number of lamellae arranged in two rows creating a large surface area.	Leaves are broad, thin and numerous, and stems and roots have a branching structure creating a large surface area compared to the volume.
Thin	The walls of the alveoli are only one cell thick.	The walls of the lamellae are only one cell thick.	Cell walls and membranes are extremely thin.
Moist	The walls of the alveoli are lined with a thin layer of moisture.	A fish lives in water containing dissolved oxygen.	All the cells are covered with a thin layer of moisture.
Transport system between the surface and body cells	A network of capillaries surrounds each alveolus.	A network of capillaries is present inside each lamella.	Direct diffusion occurs between the air and the cells due to the large surface area to volume ratio.

The effects of smoking

Cigarettes

When smoking **cigarettes** made from tobacco, **smoke** containing several thousand different chemicals including **nicotine**, **tar** and **carbon monoxide** is inhaled into the respiratory system. Cigarette smoking poses serious **health risks** including:

- **Nicotine addiction**

  Nicotine is an **addictive** substance that leads to more smoking and makes it extremely difficult to stop. Nicotine is also toxic; continued inhalation decreases appetite, increases heart rate and blood pressure, and increases the risk of a stroke and osteoporosis.

- **Reduced oxygen carrying capacity of the blood**

  Carbon monoxide combines more readily with haemoglobin than oxygen does which reduces the amount of oxygen carried to body cells. This reduces respiration and the smoker's ability to exercise. In a pregnant woman, it deprives the foetus of oxygen, reducing growth and development.

Some components of cigarette smoke

- **A persistent cough**

  Cigarette smoke causes **mucus** production to increase and the **cilia** to stop beating. Persistent coughing occurs to try and remove the extra mucus.

- **Chronic bronchitis**

  Chronic bronchitis develops when cigarette smoke irritates and inflames the walls of the **bronchi** and **bronchioles**. This, together with the increased mucus production, causes the airways to become **obstructed** making breathing difficult.

- **Emphysema**

  Emphysema develops when cigarette smoke causes the walls of the **alveoli** to become less elastic and the walls between the alveoli to breakdown which decreases their surface area. This reduces gaseous exchange, makes exhaling difficult and causes air to remain trapped in the lungs. The bronchioles often collapse when exhaling, **obstructing** the airways, making exhaling even harder.

- **Cancer of the mouth, throat, oesophagus or lungs**

  Tar and about 60 other chemicals in cigarette smoke are **carcinogenic**. These cause **cancerous tumours** to develop in the respiratory system which replace normal, healthy tissue.

N.B. Chronic bronchitis and emphysema are two types of **chronic obstructive pulmonary disease** or **COPD**.

Marijuana

Marijuana is one of the most widely used **illegal drugs** in the world. It is usually smoked and long-term use can lead to similar health problems as tobacco smokers, mainly:

- A persistent cough.
- Bronchitis.
- Frequent acute lung infections due to marijuana reducing the body's ability to fight infection.
- Lung cancer.
- Marijuana addiction.

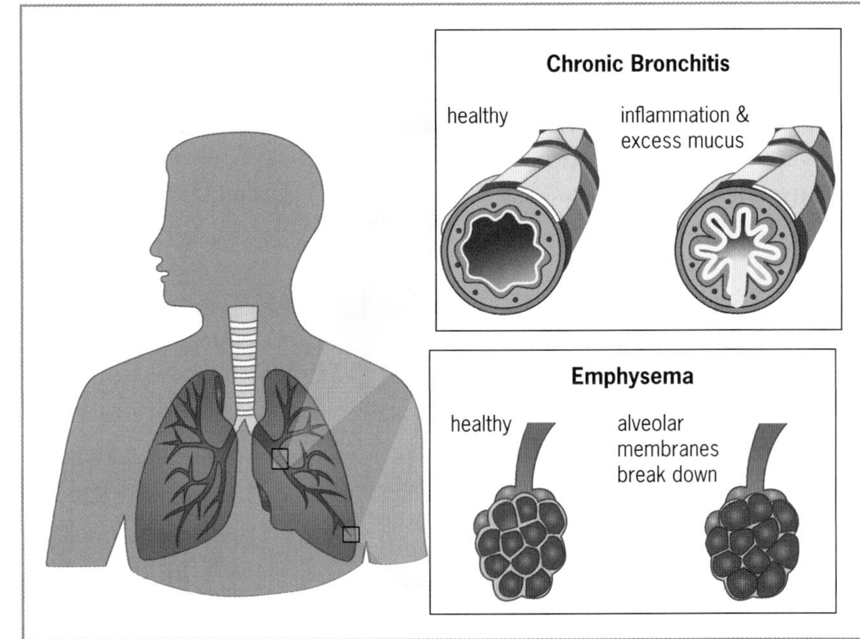

Chronic Obstructive Pulmonary Disease (COPD)

Revision questions

7. Explain the terms 'gaseous exchange' and 'breathing'.
8. Identify FOUR characteristics that gaseous exchange surfaces have in common.
9. Identify the gaseous exchange surface in a human.
10. Explain how air is drawn into the lungs in a human.
11. Explain the process of gaseous exchange in the leaves of a plant over a 24-hour period.
12. Identify the gaseous exchange surface in a fish and explain how it is adapted to perform its function efficiently.
13. You advise a friend of yours who smokes 25 cigarettes a day to stop smoking. Give FOUR reasons you could put forward to support your advice.

9 Transport systems

Living organisms need to constantly exchange substances with their environment. They need to take in useful substances and get rid of waste. **Transport systems** provide a means by which these substances are moved between the exchange surfaces and body cells.

Transport systems in multicellular organisms

The absorption and transport of substances in living organisms is affected by **two** factors:

- the **surface area to volume ratio** of the organism
- the **limitations of simple diffusion**.

Unicellular organisms, e.g. an amoeba, are very small and have a **large** surface area to volume ratio. Diffusion through their body surface is adequate to take in their requirement, e.g. oxygen, and remove their waste, e.g. carbon dioxide. In addition, no part of their body is far from its surface and substances can move these short distances by **diffusion**. They do not need a transport system to carry substances around their bodies.

Large **multicellular organisms** have a **small** surface area to volume ratio. Diffusion through their body surface is not adequate to supply all their body cells with their requirements and remove their waste. In addition, most of their body is too far from its surface for substances to move through it by diffusion. These organisms have developed **transport systems** to carry **useful substances** from specialised organs that absorb them, e.g. the lungs and ileum, to body cells, and to carry **waste substances** from body cells to specialised organs that excrete them, e.g. the kidneys.

Transport systems can carry:

- **Useful substances:** oxygen, water, digested food, e.g. glucose and amino acids, vitamins, minerals, hormones, antibodies, plasma proteins and heat in animals; and manufactured food, water and mineral salts in plants.
- **Waste substances:** carbon dioxide and nitrogenous waste, e.g. urea, in animals.

The circulatory system in humans

The **circulatory system** consists of the following:

- **Blood** which serves as the **medium** to transport substances around the body.
- **Blood vessels** which are **tubes** through which the blood flows to and from all parts of the body.
- The **heart** which **pumps** the blood through the blood vessels.

Blood

Blood is composed of **three** types of cells:

- **red blood cells**
- **white blood cells**
- **platelets.**

These cells are suspended in a fluid called **plasma**. The cells make up about 45% by volume of the blood and the plasma makes up about 55%.

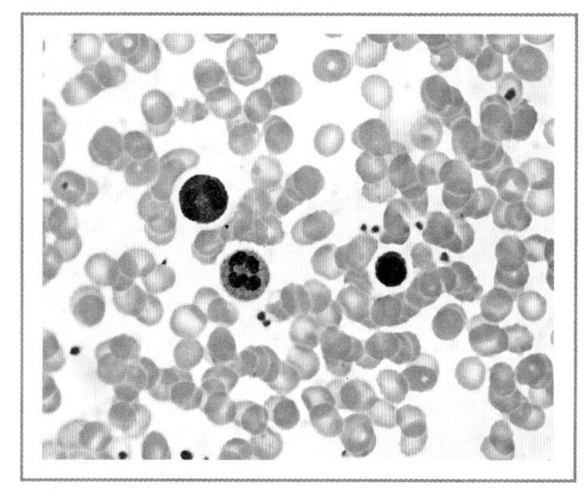

Blood cells under the microscope

Composition of plasma

Plasma is a yellowish fluid composed of about 90% **water** and 10% **dissolved substances**. The dissolved substances consist of:

- **Products of digestion**, e.g. glucose, amino acids, vitamins and minerals.
- **Waste products**, e.g. carbon dioxide as the hydrogen carbonate ion (HCO_3^-) and urea.
- **Hormones**, e.g. insulin and thyroxine.
- **Plasma proteins**, e.g. fibrinogen, albumen and antibodies.

Functions of plasma

The main function of plasma is **transporting** the following:

- **Products of digestion** from the ileum to the liver and the body cells.
- **Carbon dioxide** as the HCO_3^- ion from body cells to the lungs.
- **Urea** from the liver to the kidneys.
- **Hormones** from the glands that produce them (endocrine glands) to target organs.
- **Heat** from the liver and muscles to all parts of the body.

Blood cells

Table 9.1 *Structure and functions of blood cells*

Cell type and structure	Formation of cells	Functions
Red blood cells (erythrocytes) cell membrane cytoplasm rich in **haemoglobin**, an iron containing protein • **Biconcave discs** with a thin centre and relatively large surface area to volume ratio so gases easily diffuse in and out. • Have **no nucleus**, therefore they only live for about 3 to 4 months. • Contain the red pigment **haemoglobin**. • Slightly **elastic** allowing them to squeeze through the narrowest capillaries.	• Formed in the red bone marrow found in flat bones, e.g. the pelvis, scapula, ribs, sternum, cranium and vertebrae; and in the ends of long bones, e.g. the humerus and femur. • Broken down mainly in the liver and spleen.	• Transport **oxygen** as **oxyhaemoglobin** from the lungs to body cells. • Transport small amounts of **carbon dioxide** from body cells to the lungs.

Cell type and structure	Formation of cells	Functions
White blood cells (leucocytes) Slightly larger than red blood cells and less numerous; approximately 1 white blood cell to 600 red blood cells. There are two main types; 25% are lymphocytes and 75% are phagocytes.		
Lymphocytes • Have a **rounded** shape. • Have a large, **round nucleus** that controls the production of antibodies. • Have only a small amount of cytoplasm.	• Develop from cells in the red bone marrow and mature in other organs, e.g. lymph nodes, spleen, thymus gland.	• Produce **antibodies** to destroy disease-causing bacteria and viruses (pathogens). • Produce **antitoxins** to neutralise toxins produced by pathogens.
Phagocytes • Have a **variable** shape. • Move by **pseudopodia**; can move out of capillaries through their walls and engulf pathogens using pseudopodia. • Have a **lobed nucleus**.	• Formed in the red bone marrow.	• Engulf and destroy pathogens. • Engulf pathogens destroyed by antibodies.
Platelets (thrombocytes) • Cell **fragments**. • Have **no nucleus** and only live for about 10 days.	• Formed from cells in the red bone marrow.	• Help the blood to **clot** at a cut or wound (see page 76).

Blood groups

Blood can be classified into different **blood groups** based on chemicals present on the surface of red blood cells known as **antigens**. There are **two** grouping systems:

- The **ABO system** which divides blood into **four** groups: group A, group B, group AB and group O.
- The **rhesus system** which divides blood into **two** groups: rhesus positive and rhesus negative.

Blood and defence against disease

The blood defends the body against diseases caused by **pathogens**, mainly bacteria and viruses, in a variety of ways.

Clot formation

When the skin is cut, **platelets**, on exposure to air, release an enzyme called **thromboplastin**. Thromboplastin, with the help of calcium ions and vitamin K in the blood, starts a series of chemical reactions that finally change the soluble plasma protein called **fibrinogen** into insoluble **fibrin**. Fibrin forms a network of **fibres** across the cut that trap blood cells and form a **clot**. The clot prevents further blood loss and pathogens from entering.

The role of phagocytes

Phagocytes continuously leave the blood by squeezing between the cells of the capillary walls into tissues where they engulf and digest pathogens, especially bacteria, by **phagocytosis**.

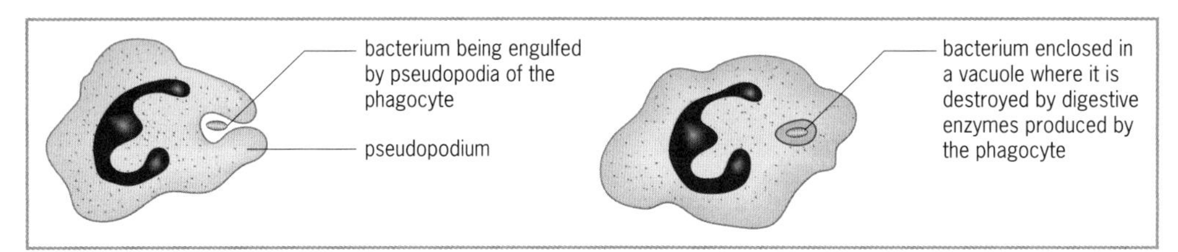

Figure 9.1 *Phagocytosis*

If tissues become **infected** by large numbers of pathogens either in a cut or wound, or inside the body, the **inflammatory response** is triggered. Blood vessels supplying the site of infection dilate and blood flow to the area increases. The response makes the area swollen and red, brings more phagocytes to the area and increases the permeability of the capillary walls. The **phagocytes** easily squeeze out of the capillaries into the tissues where they engulf and digest the pathogens.

Natural immunity

***Immunity** is the temporary or permanent resistance to a disease.*

Natural immunity results from a person having been exposed to a pathogenic disease caused by a virus or bacterium. **Lymphocytes** bring about this immunity by producing proteins called **antibodies** in response to the presence of foreign substances, known as **antigens**, in the body. Antigens include chemicals, mainly proteins, found in the walls or coats of pathogens, or toxins produced by pathogens. Antigens are **specific** to the type of pathogen and are foreign to all other organisms. When a pathogen enters the body, lymphocytes make specific antibodies in response to the pathogen's specific antigen. These **antibodies** can:

- Cause the pathogens to **clump together** so that the **phagocytes** can engulf them, or
- Cause the pathogens to **disintegrate**, or
- Neutralise the **toxins** produced by the pathogens; antibodies that do this are called **antitoxins**.

Production of antibodies takes time and the pathogen produces **symptoms** of disease before being destroyed or having its toxins neutralised. Once the person recovers, the antibodies gradually disappear from the blood and some lymphocytes develop into **lymphocyte memory cells** that remember the specific antigen.

When the pathogen enters the body again, these memory lymphocytes recognise the antigen, multiply and **quickly** produce **large quantities** of the specific antibody. The pathogen is destroyed or its toxins neutralised before symptoms of the disease develop. The person has become **immune** to the disease. Immunity may last a short time, e.g. against the common cold, to a lifetime, e.g. chicken pox is rarely caught twice.

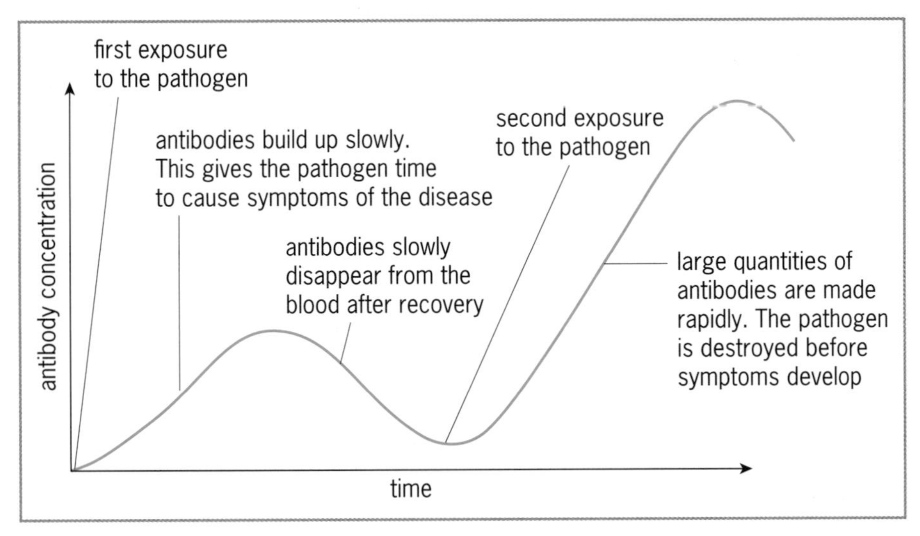

Figure 9.2 *Antibody production during the acquisition of immunity*

A **baby** gains important immunity by receiving antibodies that pass across the **placenta** before birth or from **breast milk** during breast feeding. Since the baby's lymphocytes have not been involved in producing the antibodies and the antibodies gradually disappear from the blood, immunity lasts only a short time.

Artificial immunity

Artificial immunity is acquired by **vaccination** and is used to control the spread of **communicable diseases**, i.e. diseases that pass from person to person. A **vaccine** may contain:

- Live pathogens that have been **weakened (attenuated)**, e.g. measles, mumps and rubella vaccines.
- Pathogens that have been **killed**, e.g. cholera, influenza and polio vaccines.
- **Toxins** from the pathogen that have been made harmless, e.g. diphtheria and tetanus vaccines.
- **Fragments** of the pathogen, e.g. influenza vaccine.
- The specific **antigens** (proteins) from the coat of the pathogen produced by genetic engineering, e.g. hepatitis B vaccine (see page 159).

Vaccines do not cause the disease, but lymphocytes still make **antibodies** in response to the specific **antigens** that are present in the vaccine. **Lymphocyte memory cells** are also produced so that an **immune response** is set up whenever the pathogen enters the body. Artificial immunity may last a short time, e.g. against cholera, to a lifetime, e.g. against tuberculosis.

Blood vessels

There are **three** main types of blood vessels:

- **arteries**
- **capillaries**
- **veins.**

Arteries carry blood **away** from the heart. On entering an organ, an artery branches into smaller arteries called **arterioles** which then branch into a network of **capillaries** that run throughout the organ. Capillaries then join into small veins called **venules** which join to form a single **vein** that leads back from the organ **towards** the heart.

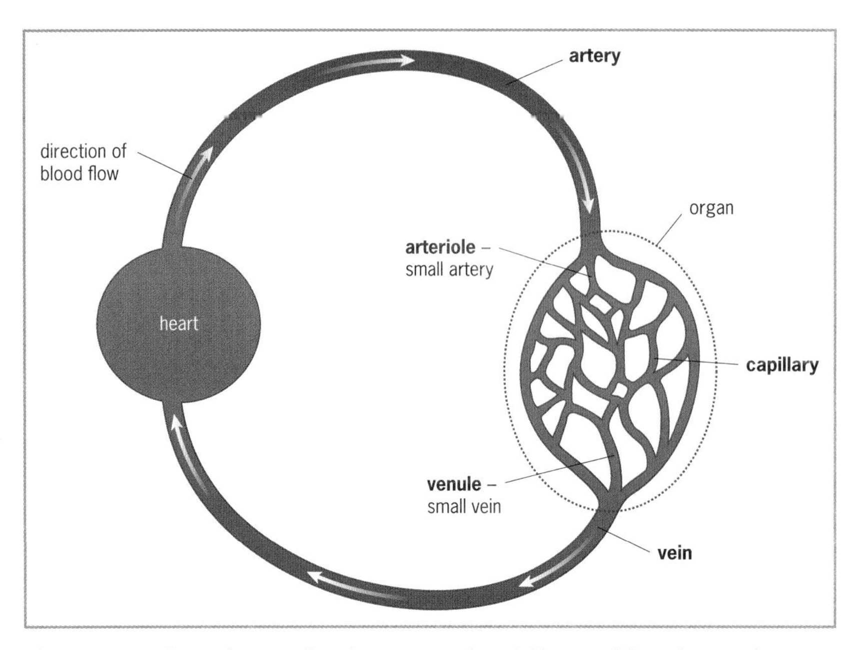

Figure 9.3 *The relationship between the different blood vessels*

Table 9.2 *Arteries, capillaries and veins compared*

Arteries	Capillaries	Veins
Transport blood **away** from the heart to body tissues and organs.	Transport blood **throughout** all body tissues and organs, linking arteries to veins.	Transport blood back **towards** the heart from body tissues and organs.
Blood flows through under **high pressure**.	Blood flows through under **low pressure**.	Blood flows through under **low pressure**.
Blood moves in **pulses** created as the ventricles contract.	Blood flows **smoothly**.	Blood flows **smoothly**.
Blood flows **rapidly**.	Blood flows **very slowly**.	Blood flows **slowly**.
Blood is **oxygenated**, except in the pulmonary artery.	Blood becomes **deoxygenated** as it travels through capillaries.	Blood is **deoxygenated**, except in the pulmonary vein.
Most lie **deep** within the body so they are protected.	Run **throughout** all tissues and organs.	Many lie **close** to the body surface.
Do not possess valves, except the aorta and pulmonary artery as they leave the ventricles of the heart.	**Do not** possess valves.	Possess **valves** to prevent the low pressure, slow flowing blood from flowing backwards. free-flowing blood – **valve open** back-flowing blood – **valve closed**

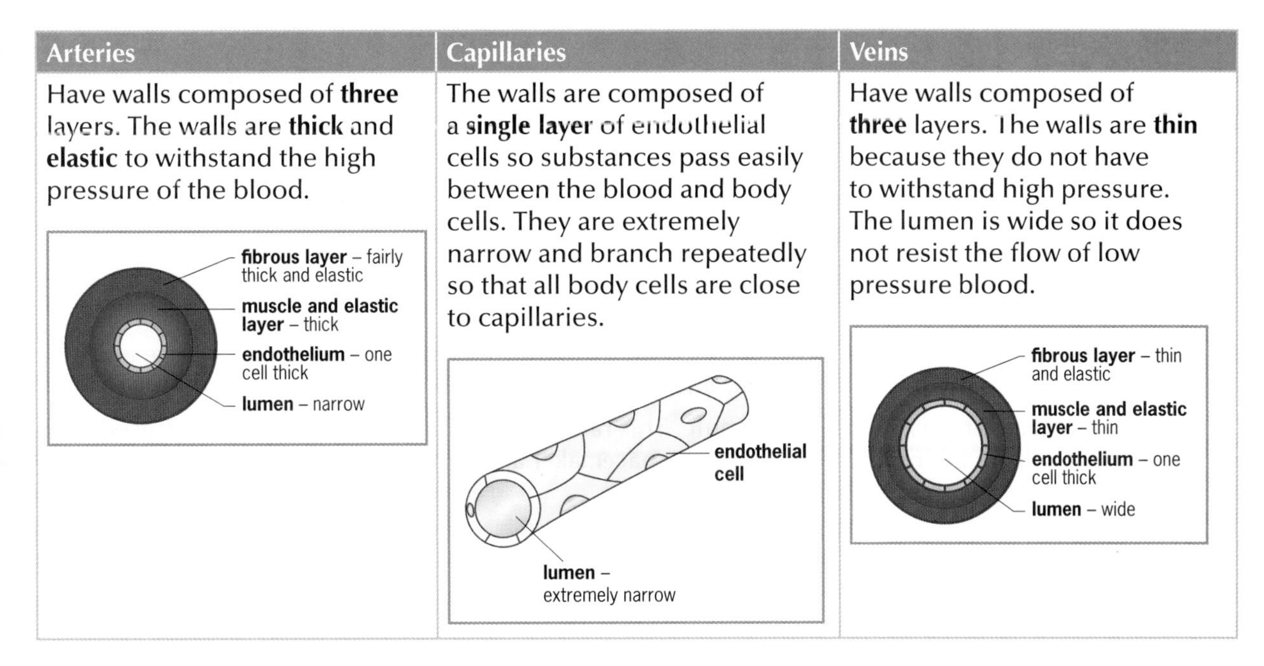

Arteries	Capillaries	Veins
Have walls composed of **three** layers. The walls are **thick** and **elastic** to withstand the high pressure of the blood.	The walls are composed of a **single layer** of endothelial cells so substances pass easily between the blood and body cells. They are extremely narrow and branch repeatedly so that all body cells are close to capillaries.	Have walls composed of **three** layers. The walls are **thin** because they do not have to withstand high pressure. The lumen is wide so it does not resist the flow of low pressure blood.

The heart

The pumping action of the **heart** maintains a constant circulation of blood around the body. The walls of the heart are composed of **cardiac muscle** which has its own **inherent rhythm** and does not get tired.

The heart is divided into **four** chambers. The two on the right contain **deoxygenated blood** and are completely separated from the two on the left which contain **oxygenated blood**, by the **septum**. The top two chambers, called **atria**, have thin walls and they collect blood entering the heart. The bottom two chambers, called **ventricles**, have thick walls and they pump blood out of the heart. **Valves** are present between each atrium and ventricle and in the pulmonary artery and aorta as they leave the ventricles to ensure that blood flows through the heart in **one direction**.

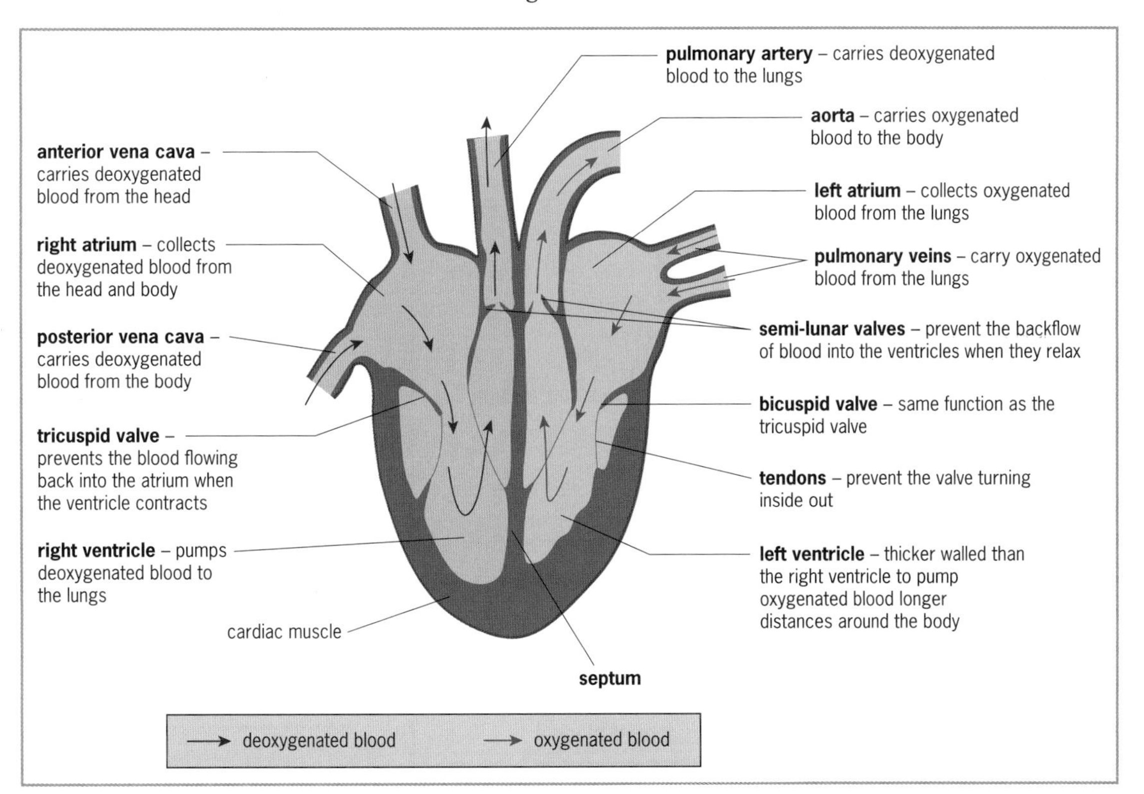

Figure 9.4 *Longitudinal section through the human heart showing the function of the parts*

Cardiac cycle

The atria and ventricles at the two sides of the heart contract and relax together. During one **cardiac cycle** or **heartbeat**:

- The **atria** and **ventricles relax** together (**diastole**), the semi-lunar valves close, the atria fill up with blood from the anterior and posterior vena cavae and pulmonary vein, and the blood flows into the ventricles. This takes 0.4 seconds.
- The **atria contract** together (**atrial systole**) forcing any remaining blood into the ventricles. This takes 0.1 seconds.
- The **ventricles contract** together (**ventricular systole**), the tricuspid and bicuspid valves close and blood is forced into the pulmonary artery and aorta. This takes 0.3 seconds.

The heart beats on average 75 times per minute. This rate is maintained by a group of specialised cardiac muscle cells in the right atrium called the **pacemaker** and can be modified by nerve impulses, e.g. the rate increases with **exercise**.

Circulation

During one complete circulation around the body, the blood flows through the heart **twice**, therefore, humans have a **double circulation**:

- In the **pulmonary circulation**, blood travels from the **right ventricle** through the **pulmonary artery** to the **lungs** to pick up oxygen and lose carbon dioxide, i.e. it becomes **oxygenated**. It then travels back via the **pulmonary vein** to the **left atrium**.
- In the **systemic (body) circulation**, the blood travels from the **left ventricle** through the **aorta** to the **body** where it gives up oxygen to the body cells and picks up carbon dioxide, i.e. it becomes **deoxygenated**. It then travels back via the **anterior** or **posterior vena cava** to the **right atrium**.

A **double circulation** is necessary because blood **loses pressure** when it passes through the lungs, so it goes back to the heart to be given enough pressure to reach body organs to supply them with oxygen. As it loses pressure passing through organs, the blood goes back to the heart again to be given enough pressure to reach the lungs to get rid of waste carbon dioxide and pick up more oxygen.

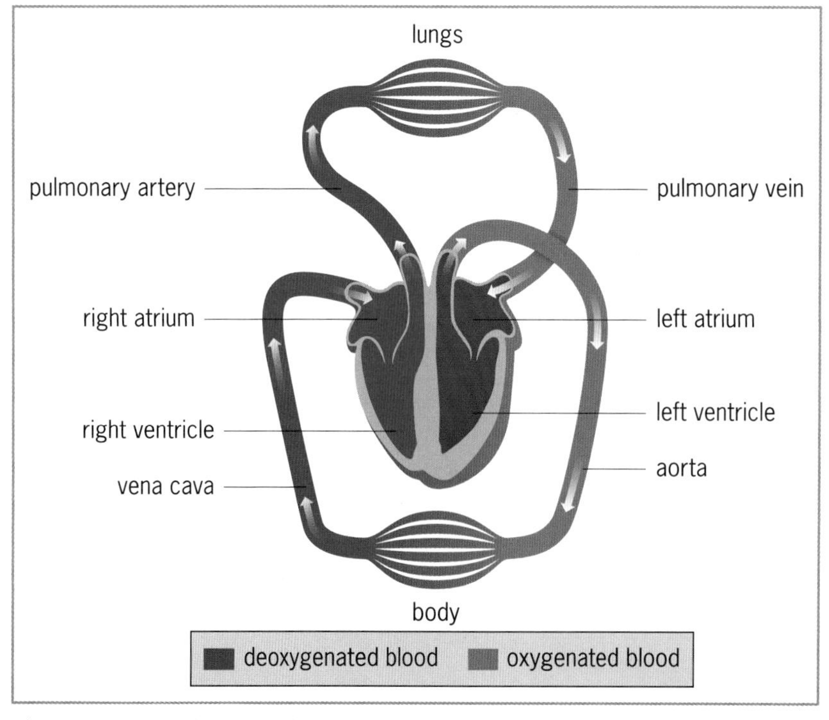

Figure 9.5 *Double circulation in the human body*

Pathway of blood around the body

The names of the blood vessels supplying the **major organs** are given in Figure 9.6. Using these it is possible to describe the pathway of blood around the body.

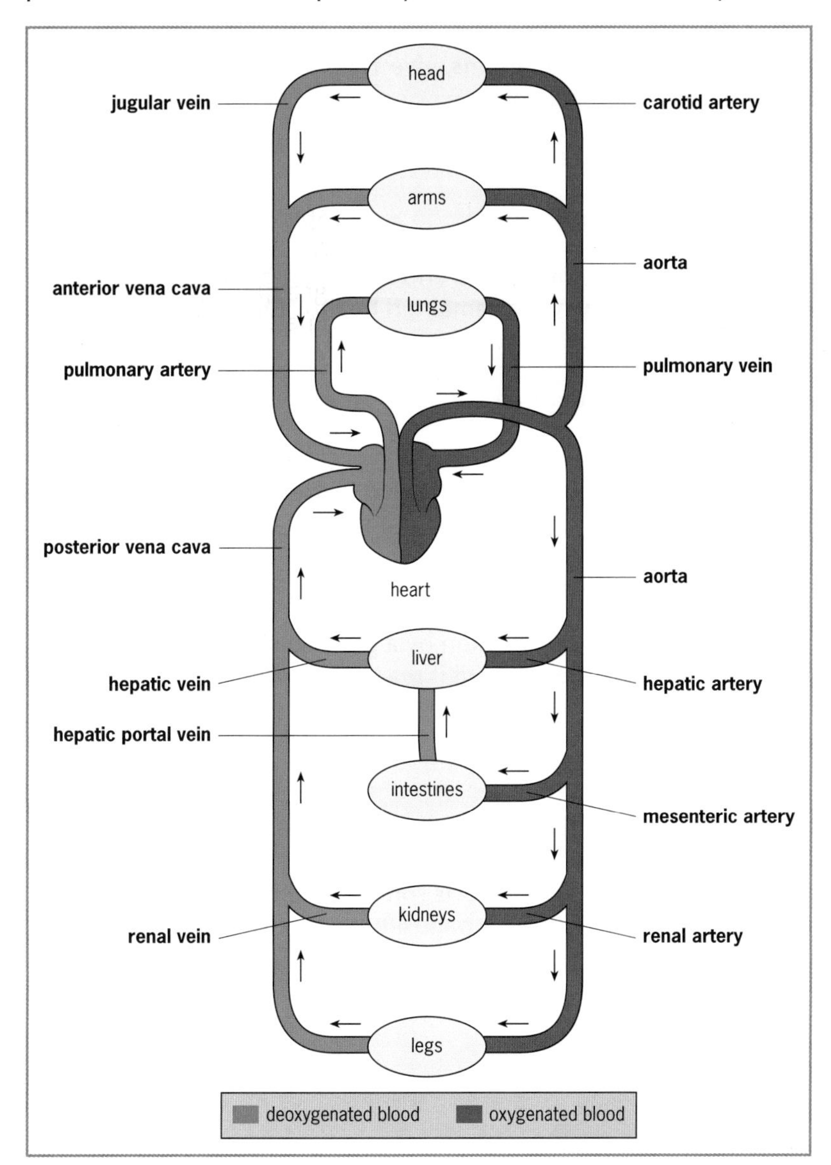

Figure 9.6 *The major blood vessels in the human body*

Example

A red blood cell starting in the **left ventricle** and returning to that chamber after passing through the **intestines** would take the following pathway:

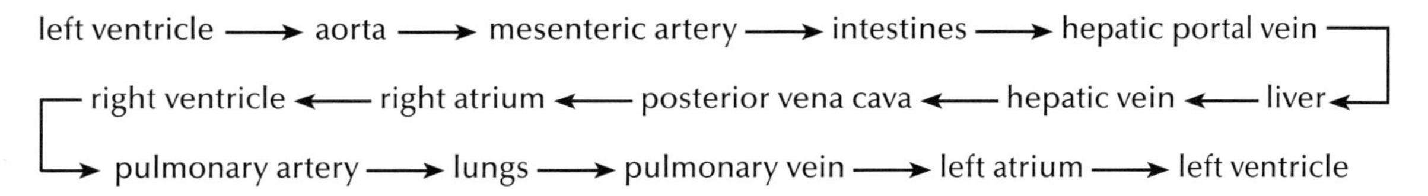

Revision questions

1. Explain why a monkey has a well developed transport system, but an amoeba lacks any form of a transport system in its body.
2. By means of TWO labelled and annotated diagrams, give THREE differences between the structure of a red blood cell and a phagocyte.
3. State the function of a red blood cell and a phagocyte.
4. Explain how the loss of blood at a cut is prevented.
5. Alicia contracted chicken pox when she was a child and her son John recently contracted the disease. Explain fully how Alicia remained healthy despite her close contact with John throughout his illness.
6. State THREE differences between the structure of an artery and a vein, and provide a reason for EACH difference.
7. Name the tissue that makes up the walls of the heart. What is special about this tissue?
8. Explain how the flow of blood through the heart is maintained in one direction.
9. Draw a simple flow diagram to show the route taken by a glucose molecule from the time it is absorbed into the blood in the ileum until it reaches the brain.

Transport systems in flowering plants

Substances are transported around plants by **vascular tissue** composed of **xylem** and **phloem tissues**. Xylem tissue, composed of xylem vessels, transports **water** and **mineral salts**. Phloem tissue, composed of phloem sieve tubes and companion cells, transports **soluble food materials**. Vascular tissue runs throughout roots, stems and leaves (see Figure 7.3, page 48).

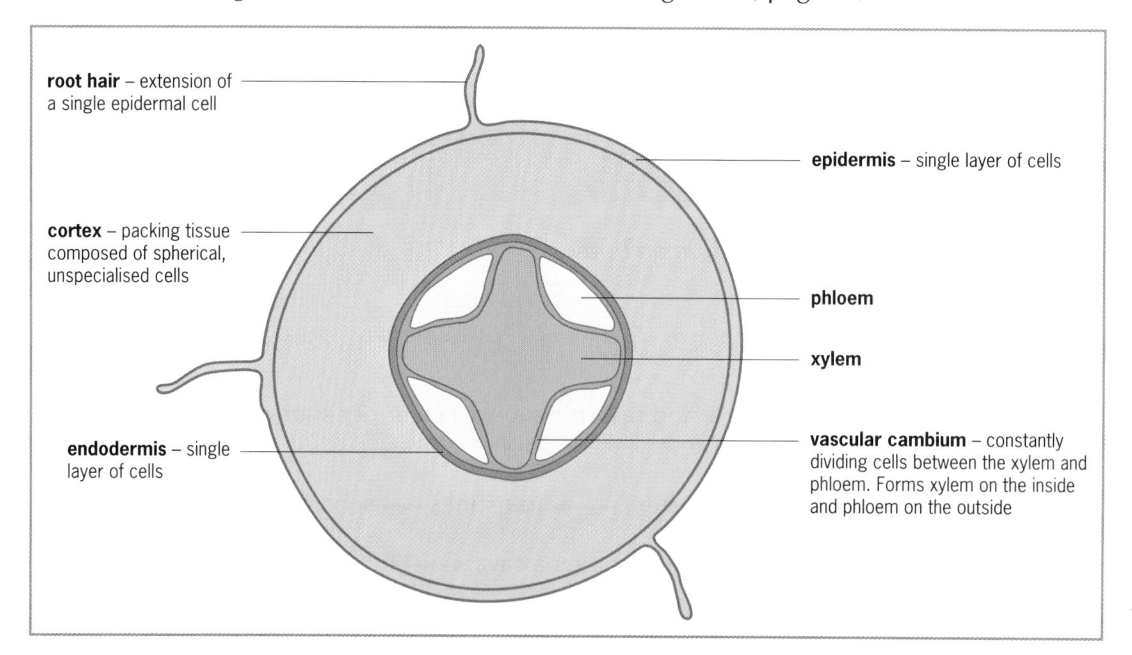

Figure 9.7 *Transverse section through a root of a dicotyledon showing the arrangement of xylem and phloem*

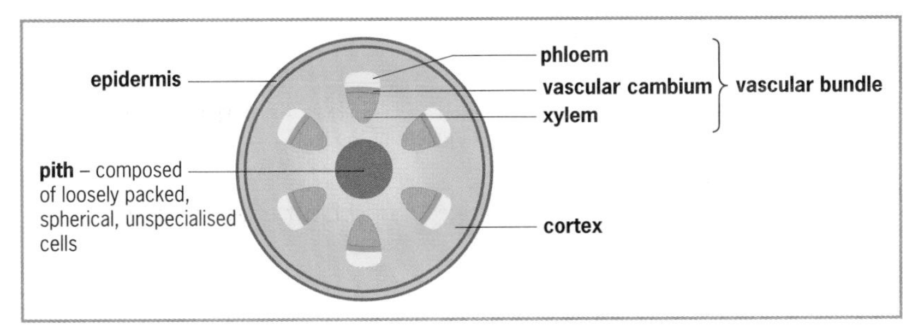

Figure 9.8 *Transverse section through a young stem of a dicotyledon*

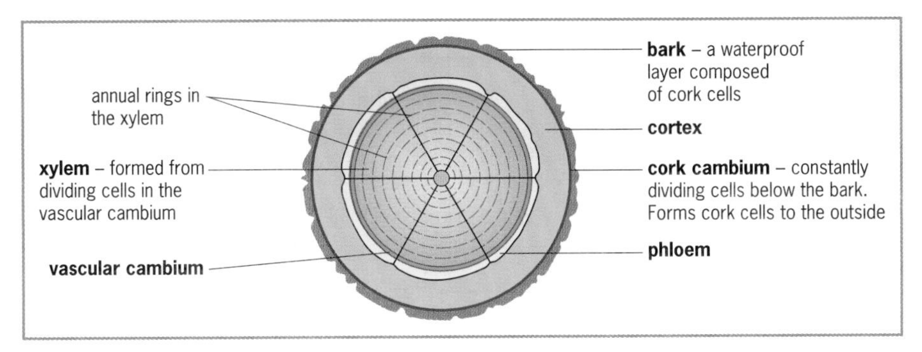

Figure 9.9 *Transverse section through a woody stem of a dicotyledon*

Movement of water through a flowering plant

Xylem vessels transport **water** from roots to leaves for use in photosynthesis. Xylem vessels are long, narrow, hollow tubes that are non-living and are formed from columns of elongated cells. The contents of the cells die, the cross walls between adjacent cells in each column breakdown and the cellulose walls of the vessels become thickened with **lignin** in rings, spirals or a net-like pattern. Being **long**, **narrow** and **hollow** with **no cross walls**, water can flow continuously through xylem vessels. The lignin is tough and strong so xylem vessels also help **support** the plant. **Wood** is almost entirely composed of lignified xylem vessels.

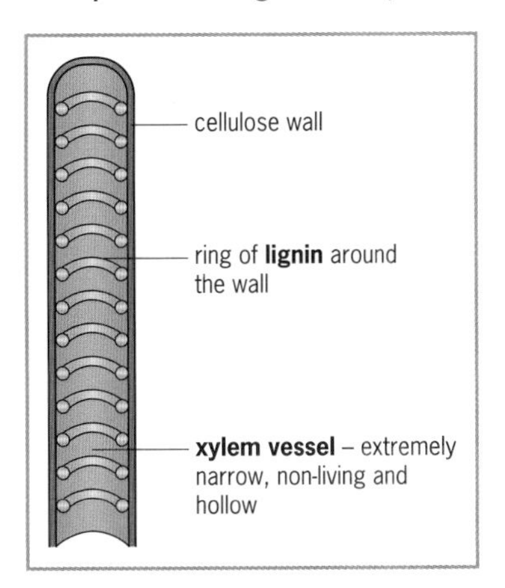

Figure 9.10 *Longitudinal section through a xylem vessel*

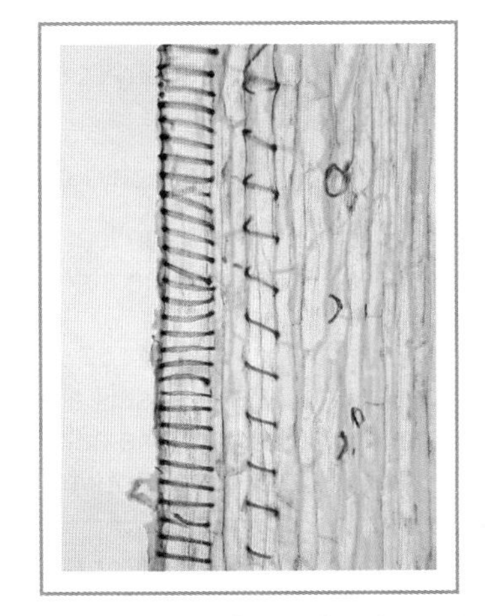

Xylem vessels under the microscope

Water moves through a flowering plant by a combination of:

- **root pressure**
- **transpiration**
- **capillarity.**

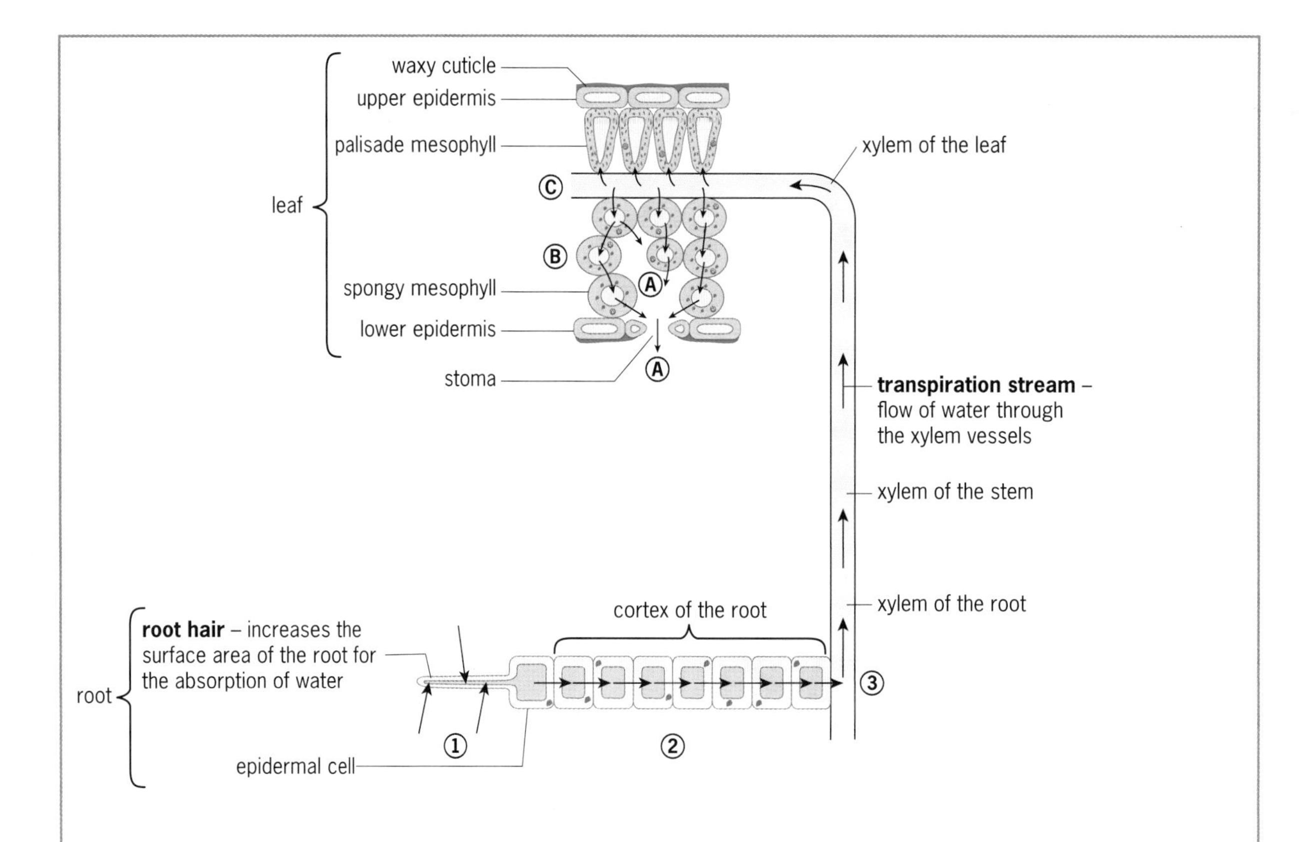

- **Root pressure** is a **push** from the roots, i.e. ① to ③:

  ① Water enters the root hairs and epidermal cells by **osmosis**.

  ② Water moves through the cortex cells of the root by **osmosis**.

  ③ Water is pushed into the xylem of the root. This push is created by ① and ② and is called **root pressure**. Root pressure pushes the water up the xylem in the root and into the bottom of the xylem in the stem.

- **Transpiration** is the loss of water vapour from the surface of leaves which creates a **pull**, i.e. Ⓐ to Ⓒ:

  Ⓐ Water **evaporates** from the spongy mesophyll cells around the air spaces and the water vapour **diffuses** out through the stomata.

  Ⓑ Water is drawn through other spongy mesophyll cells by **osmosis**.

  Ⓒ Water is drawn from the xylem vessels in the leaf by **osmosis** and the pull created by Ⓐ and Ⓑ. This draws water up the xylem vessels in the stem. Some of the water also enters the palisade cells for use in photosynthesis.

- **Capillarity** helps water move. The xylem vessels are extremely narrow and act like **capillary tubes**. The cohesion between water molecules and the adhesion of the molecules to the xylem walls helps maintain a continuous column of water from roots to leaves and helps the water move upwards.

Figure 9.11 *Mechanism of the movement of water through a flowering plant*

Control of the loss of water by stomata

Stomata, found mainly in the lower surface of leaves, control water loss from leaves. Stomata **open** when the guard cells are **turgid** which results in **rapid** transpiration, and **almost close** when the guard cells are **flaccid** resulting in **slow** transpiration.

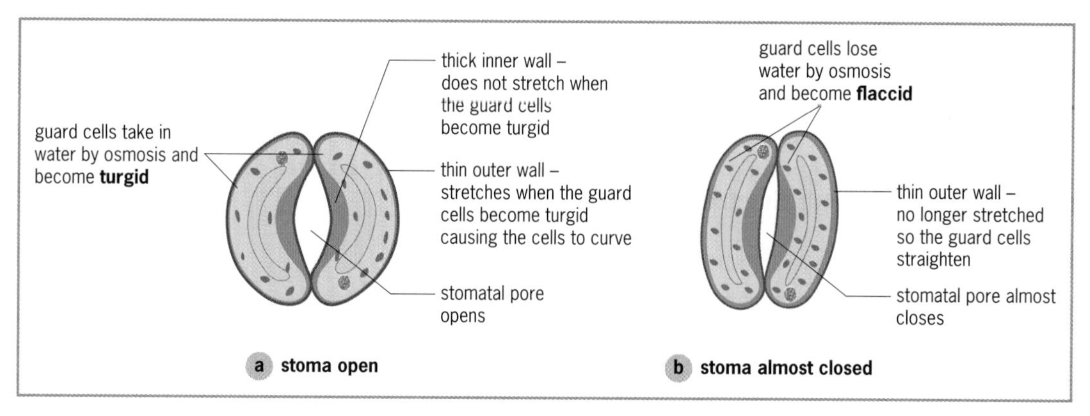

Figure 9.12 *Opening and closing of stomata*

If the supply of water in the soil is **low**, plant cells become **flaccid** so stomata almost close. This reduces water loss by transpiration and conserves water. A plant will **wilt** when it loses more water in transpiration than its roots can take up.

Environmental conditions affecting the rate of transpiration

If the water supply in the soil is **plentiful**, the **rate of transpiration** is controlled by the interaction of different **environmental conditions** that affect the rate of **evaporation** and **diffusion**.

- **Temperature**

  High temperatures cause water to evaporate and water vapour to diffuse rapidly so transpiration is **rapid**. **Low temperatures** cause water to evaporate and water vapour to diffuse slowly so transpiration is **slow**.

- **Humidity**

  In **low humidity**, the concentration gradient between the water vapour in the air spaces in leaves and the air surrounding the leaves is high, so water vapour diffuses out easily and transpiration is **rapid**. In **high humidity**, the air surrounding the leaves is almost saturated with water vapour so the concentration gradient is low and very little more can diffuse out causing transpiration to be **slow**.

- **Wind speed**

  In **windy conditions**, water vapour is carried away from the surface of leaves so more can diffuse out and transpiration is **rapid**. In **still conditions**, water vapour remains around the leaves and very little more can diffuse out so transpiration is **slow**.

- **Light intensity**

  In **bright light**, the stomata are fully open so water vapour can diffuse out easily and transpiration is **rapid**. In **dim light**, the stomata almost close so little water vapour can diffuse out and transpiration is **slow**.

The importance of transpiration

Transpiration is important to plants for the following reasons:

- It draws water up to leaves for use in **photosynthesis**.
- It supplies plant cells with water to keep them **turgid**. This supports non-woody stems and leaves.
- Moving water carries dissolved **mineral salts** up to the leaves.
- Evaporation of water from the surface of leaves **cools** the plant.

Water conservation in plants

In addition to stomata almost closing when water supplies in the soil are low, terrestrial plants have developed other **adaptations** to help conserve water, especially those living in dry regions (xerophytes), e.g. deserts, or in regions of high salinity (halophytes), e.g. coastal regions. Many adaptations conserve water by **reducing transpiration**.

Table 9.3 *Methods by which plants conserve water*

Method	Adaptations
Reducing the rate of transpiration	• Leaves have extra-thick waxy cuticles. • Leaves have reduced numbers of stomata. • Stomata are grouped together in sunken pits that trap water vapour, e.g. oleander. • Stomata almost close in the daytime if temperatures are very high and open at night. • Leaves can roll with the stomata to the inside, e.g. marram grass. • Leaves have fine hairs on their surface that trap water vapour. • The surface area of leaves is reduced, e.g. needle-shaped leaves of conifers, spines of cacti and scales of casuarina. • Leaves are shed in the dry season or winter months.
Storing water	• Leaves of many succulent plants store water, e.g. aloe. • Stems of many succulent plants store water, e.g. cacti. The baobab tree stores water in its trunk. • Roots of some plants store water, e.g. some members of the pumpkin family.
Increasing the uptake of water	• Plants have very long, deep tap-roots to absorb water from deep in the soil. • Plants have shallow, widespread root systems to absorb surface water from a wide area.

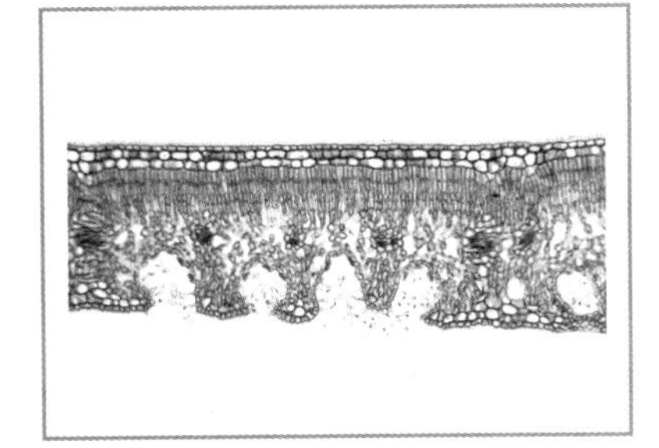

Stomata in sunken pits

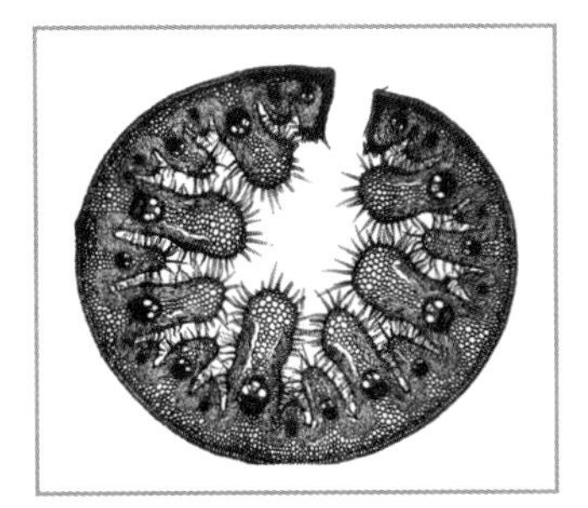

Rolled leaf of marram grass

Succulent stems and needle-shaped leaves of cacti

Movement of organic food through a flowering plant

Phloem sieve tubes transport organic food, mainly **sucrose** and some **amino acids**, from leaves to all other parts of the plant. These are long, narrow tubes that are formed from columns of elongated cells. The cross walls between adjacent cells in each column become perforated by small holes to form **sieve plates**. Each cell is called a **sieve tube element** and it contains living cytoplasm but no nucleus. The cytoplasm of adjacent sieve tube elements is connected through the holes. Each sieve tube element has a **companion cell** next to it that contains a nucleus. The nucleus controls the functioning of both cells.

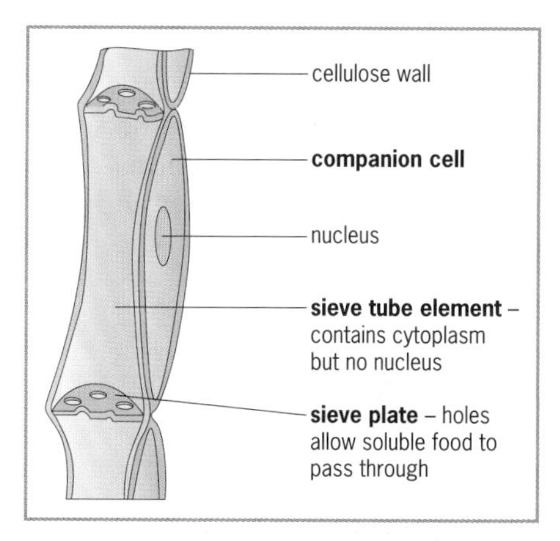

Figure 9.13 *Longitudinal section through a phloem sieve tube and companion cell*

The movement of organic food through phloem sieve tubes is called **translocation**.

Mechanism of translocation

The **pressure flow hypothesis** helps to explain how **dissolved sugars** move through the phloem. The hypothesis states that sugars flow from a **sugar source** to a **sugar sink**. During the process, a **pressure gradient** is created between the source and the sink which causes the contents of the phloem sieve tubes to move both **upwards** and **downwards**.

- **Sugar sources** include parts of plants that **produce sugars**, e.g. photosynthesising leaves, or parts that **release sugars**, e.g. storage organs at the beginning of the growing season and the cotyledons of seeds at the beginning of germination where stored starch is hydrolysed to sugars.
- **Sugar sinks** are parts of plants that **require sugars** including stems, roots, fruits, storage organs and growing parts. Sugars in sugar sinks may be used in respiration, stored or converted to other substances which are either stored, e.g. starch, or used in growth, e.g. proteins.

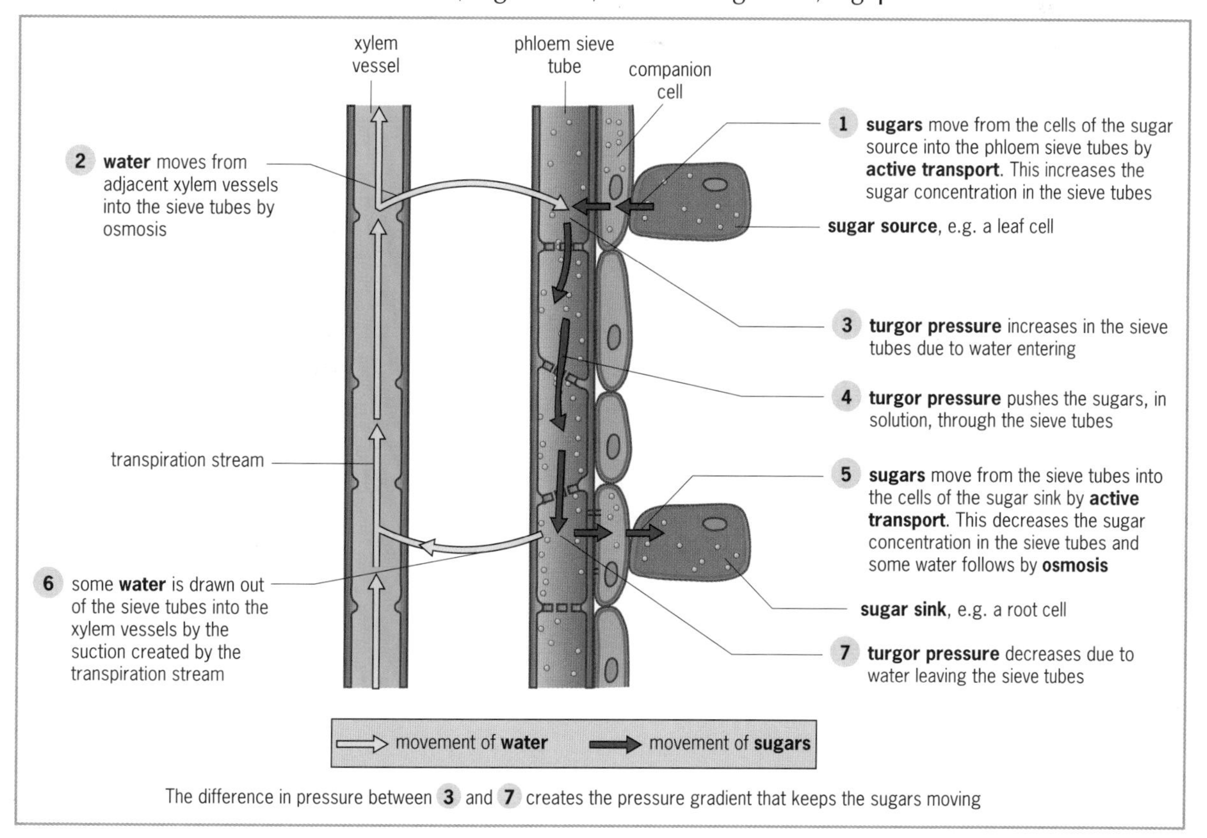

Figure 9.14 *The mechanism of translocation*

Movement of mineral salts through a flowering plant

Mineral salts are absorbed by roots in the form of **ions**. The ions are dissolved in the water in the soil and are absorbed by the root hairs against a concentration gradient by **active transport**. The ions then move through the cortex cells and into the xylem vessels dissolved in the moving water and are carried throughout the plant dissolved in the water in the **xylem vessels**.

Storage of food in living organisms

The importance of food storage in living organisms

Food that is not needed immediately is stored in living organisms for future use. Food storage is **important**:

- It overcomes the need for continuous food intake in heterotrophs and continuous food manufacture in autotrophs.
- It provides **food reserves** for periods of food scarcity in heterotrophs, e.g. during the winter in temperate climates, and periods when food manufacture cannot occur in autotrophs, e.g. during a drought or when temperatures are very low. This enables animals to hibernate and plants to survive through the dry season or winter months.
- To provide food reserves for **special functions**, e.g. the production of sexual or vegetative reproductive structures and the development of embryos. The seeds of plants and eggs of many animals store food for use by the embryos as they grow, and new plants use stored food as they develop from vegetative organs.

Soluble food substances, e.g. glucose, amino acids, fatty acids and glycerol, are usually **condensed** to **insoluble substances**, e.g. starch, glycogen, protein and lipid, to be stored. These insoluble substances do not interfere with osmosis and other cellular processes and can be **hydrolysed** to soluble substances when required.

Storage of food in animals

Animals store mainly **glycogen** and **fat**. They do not store protein.

- **Storage in adipose tissue**. Excess fat is stored in fat cells found adipose tissue under the skin and around organs. Excess glucose can also be converted to fat and stored.
- **Storage in the liver**. The liver stores:
  - **Glycogen** which is formed by condensation of excess glucose in the blood.
  - **Vitamins** A, B_{12} and D.
  - **Iron** which is formed from the breakdown of haemoglobin in red blood cells.
- **Storage in skeletal muscle**. Skeletal muscles store **glycogen** formed by condensation of excess glucose in the blood. The muscle cells can then convert this back to glucose for use in respiration to provide energy when necessary, e.g. during exercise.

Storage of food in plants

Food can be stored in **roots**, **stems**, **leaves**, **fruits** and **seeds**. The phloem sieve tubes transport **sugars**, made in photosynthesis, to these structures where they are stored, or are converted to starch, oils or proteins and stored.

- **Storage in vegetative organs**. Vegetative organs are **underground structures** that are swollen with food at the end of the growing season. They allow the plant to survive through the unfavourable season, e.g. the dry season or winter, and to grow rapidly, using stored food, at the beginning of the favourable season, e.g. the rainy season or summer. They can also act as a means of **asexual reproduction** since several new plants can grow from one organ. Vegetative organs can be:
  - **Stems** such as stem tubers, e.g. yam and English potato, rhizomes, e.g. ginger, and corms, e.g. eddo.
  - **Roots** such as root tubers, e.g. sweet potato.
  - **Leaves** or **leaf bases** such as those that make up bulbs, e.g. onion.

  The main food stored in most is **starch**.
- **Storage in tap-roots**. Tap-roots are single, vertical roots. They store **starch**, e.g. turnip, or **sugars**, e.g. carrot and sugar beet.
- **Storage in succulent fruits**. Succulent fruits store mainly **sugars**, e.g. mango and paw-paw. Some store **starch**, e.g. breadfruit. The fruits of avocado and olive store **oil**. The stored food attracts animals to eat the fruits and this helps disperse the seeds (see page 131).
- **Storage in seeds**. The **cotyledons** and **endosperm** of seeds can store **starch**, e.g. rice and wheat, **protein**, e.g. peas and beans, and **oil**, e.g. nuts. This stored food is then used when the seeds germinate (see page 114).
- **Storage in stems**. The stems of **sugar cane** store **sucrose** in the vacuoles of their cells. The stems of some **succulent** plants store **water** in their cells, e.g. cacti.
- **Storage in leaves**. The leaves of some **succulent** plants store **water** in their cells, e.g. aloe.

A rhizome

Revision questions

10 State the function of the xylem vessels of a plant and explain how their structure is suited to their function.

11 Explain the mechanism by which water from the soil reaches the leaves of a plant.

12 Account for the fact that a plant absorbs more water on a hot, windy day than on a cool, still day.

13 Outline the role played by transpiration in the life of a flowering plant.

14 Plants living in regions where water supplies in the soil are low develop various adaptations to aid in their survival, suggest FOUR of these adaptations.

15 What is translocation?

16 Explain how sugars produced in the leaves of a mango plant reach the fruits of the plant.

17 Give TWO reasons why it is important for living organisms to store food.

18 Outline the role played by EACH of the following in food storage:

a the liver in humans **b** vegetative organs of plants.

10 Excretion and osmoregulation

Chemical reactions occurring in living organisms constantly produce **waste** and **harmful** substances which the organisms must get rid of from their bodies.

Excretion

__Excretion__ is the process by which waste and harmful substances, produced by chemical reactions occurring inside body cells, i.e. the body's metabolism, are removed from the body.

Excretion is **important** in living organisms because many waste products are **harmful** and if these build up in cells, they damage and kill the cells.

Excretion must not be confused with **egestion** which is the removal of undigested dietary fibre and other materials from the body as faeces. This dietary fibre is not produced in the body's metabolism, so its removal cannot be classed as excretion.

Products excreted by plants

Plants produce the following waste substances during metabolism:

- **Oxygen** is produced in photosynthesis and is excreted during the day when the rate of photosynthesis is higher than the rate of respiration.
- **Carbon dioxide** is produced in respiration and is excreted during the night when no photosynthesis is occurring.
- **Water** is produced in respiration and is excreted during the night when no photosynthesis is occurring.
- **Organic waste products** such as tannins, alkaloids, anthocyanins, and salts of organic acids such as **calcium oxalate**.

Mechanisms of excretion in plants

Plants, unlike animals, do not have any specialised excretory organs.

- **Oxygen, carbon dioxide** and **water vapour** diffuse out through the **stomata** of leaves and **lenticels** of bark-covered stems and roots.
- **Organic waste products** can be stored in dead, permanent tissue, e.g. heart wood. They can also be converted to **insoluble** substances such as oils or insoluble crystals, e.g. excess calcium ions combine with the waste product, oxalic acid, to form calcium oxalate crystals. In this insoluble form, they do not affect osmotic and metabolic processes in cells and can be stored in the cells of leaves, bark, petals, fruits and seeds. They are then removed when the plant **sheds** these structures.

Products excreted by animals

Animals produce the following waste substances during metabolism:

- **Carbon dioxide** is produced in respiration.
- **Water** is produced in respiration.
- **Nitrogenous compounds** are produced by the **deamination** of amino acids in the liver, e.g. urea, ammonia which is very toxic, and uric acid which is the least toxic.
- **Bile pigments**, e.g. bilirubin, are produced by the breakdown of haemoglobin from red blood cells in the liver.
- **Heat** is produced in general metabolism.

Excretory organs in humans

Humans have several organs that excrete waste products.

- The **kidneys** excrete water, nitrogenous waste (mainly urea) and salts as **urine**.
- The **lungs** excrete carbon dioxide and water vapour during exhalation (see page 68).
- The **skin** excretes water, urea and salts as **sweat**. It also excretes heat (see page 111).
- The **liver** excretes bile pigments. It also makes nitrogenous waste.

The kidneys and excretion in humans

Humans have two **kidneys** that form part of the **urinary system**. Each kidney is divided into three regions: an outer region called the **cortex**, an inner region called the **medulla**, and a central hollow region called the **pelvis**. A **renal artery** carries blood to each kidney and a **renal vein** carries blood away.

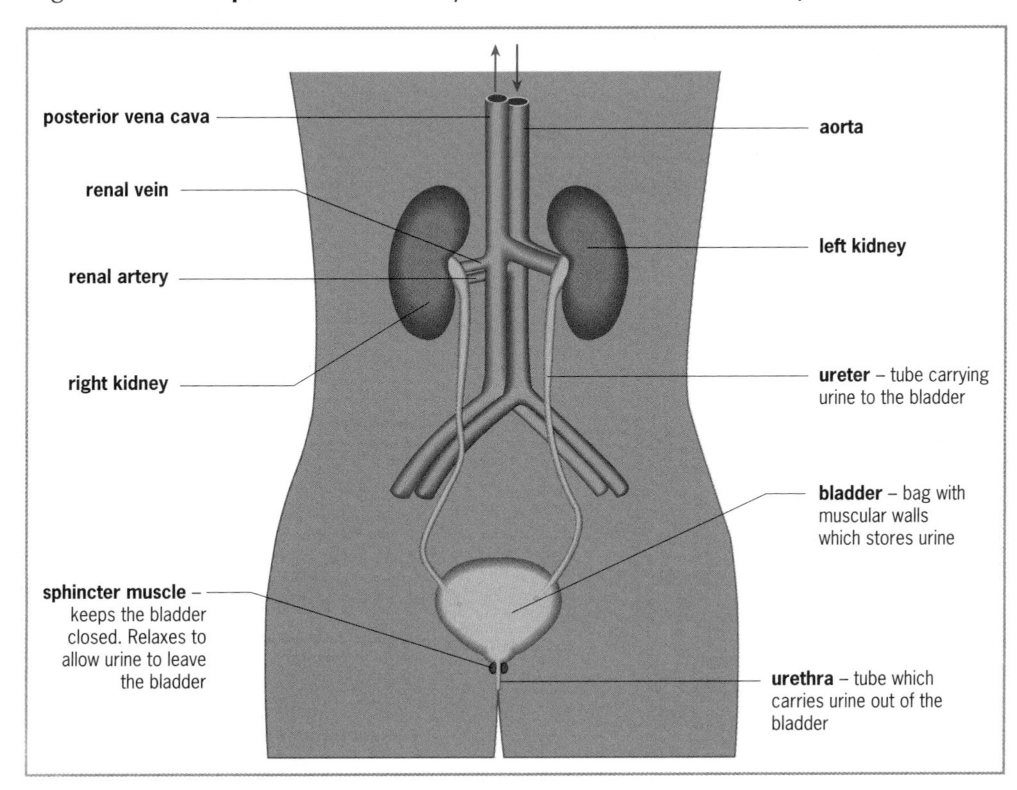

Figure 10.1 *Structure of the urinary system in a human*

Each kidney is composed of thousands of **kidney tubules** or **nephrons** that produce urine. Each nephron begins with a cup-shaped **Bowman's capsule** in the cortex which surrounds an intertwined cluster of capillaries called a **glomerulus**. After the Bowman's capsule, each nephron is divided into **three** sections:

- The **first convoluted (coiled) tubule** in the cortex.
- The **loop of Henle** in the medulla.
- The **second convoluted (coiled) tubule** in the cortex.

An arteriole, which branches from the renal artery, leads into each glomerulus. Each nephron has a network of **blood capillaries** wrapped around it which leads from the glomerulus and joins into a venule which leads into the renal vein. Nephrons join into **collecting ducts** in the cortex and these ducts lead through the medulla and out into the pelvis.

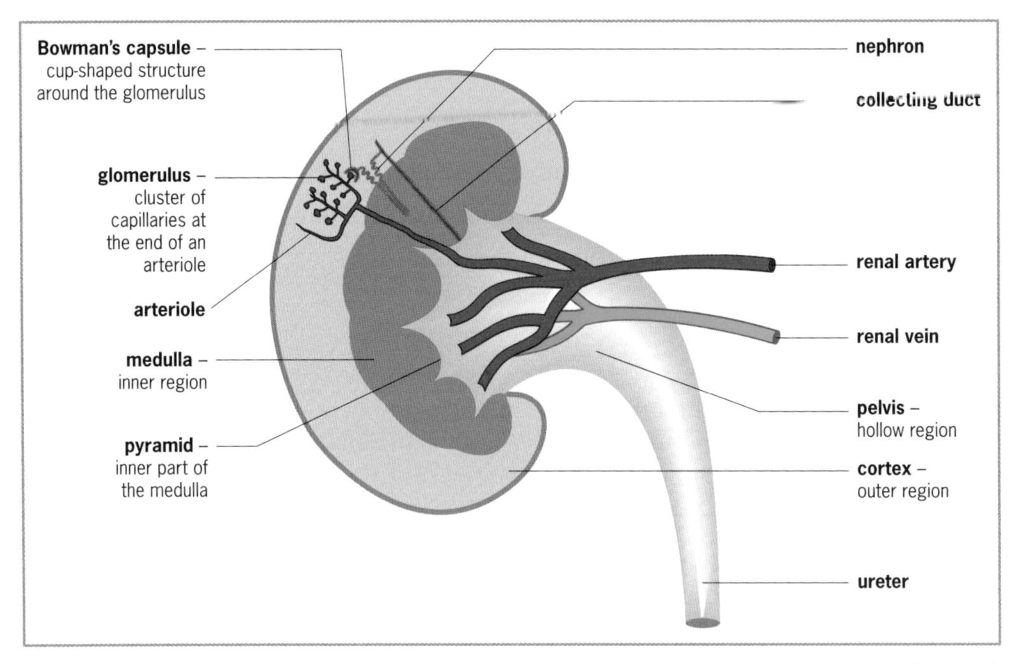

Figure 10.2 *A longitudinal section through a kidney showing the position of a nephron*

Urine is produced in the nephrons by **two** processes:

- **ultra-filtration or pressure filtration**
- **selective reabsorption.**

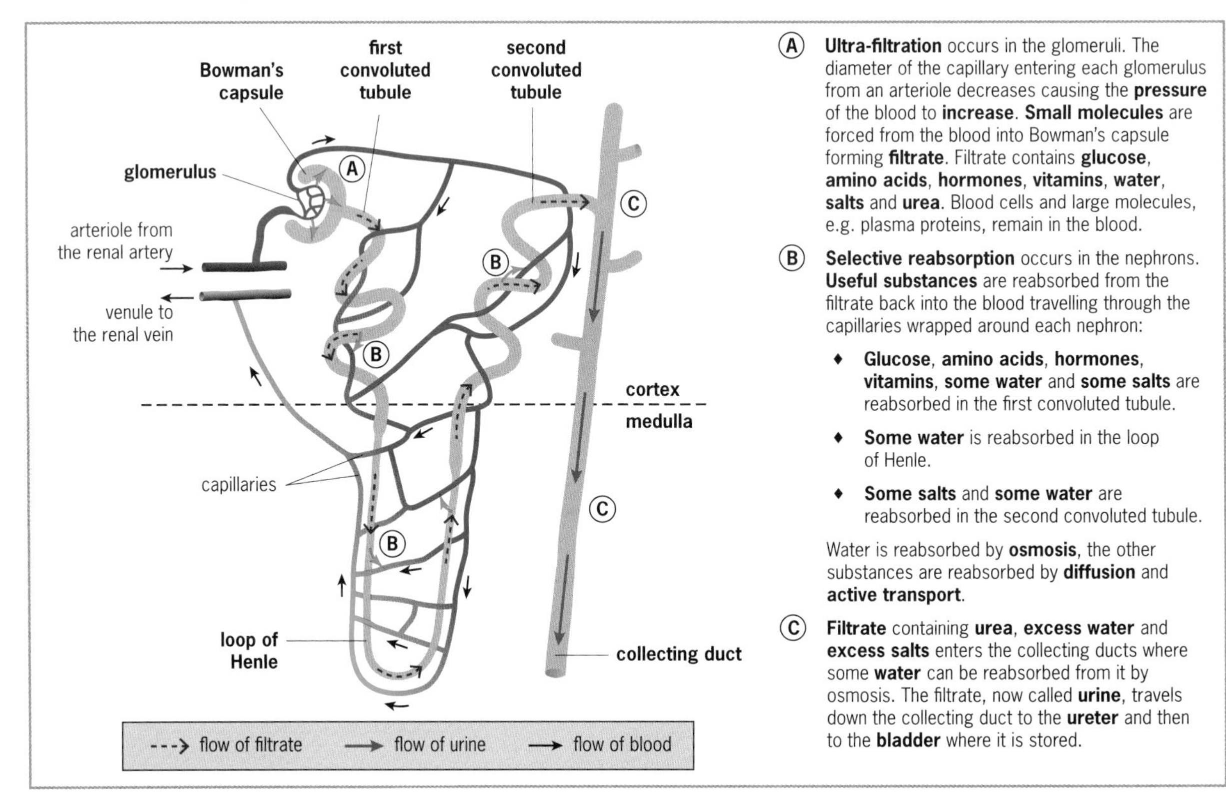

Ⓐ **Ultra-filtration** occurs in the glomeruli. The diameter of the capillary entering each glomerulus from an arteriole decreases causing the **pressure** of the blood to **increase**. **Small molecules** are forced from the blood into Bowman's capsule forming **filtrate**. Filtrate contains **glucose**, **amino acids**, **hormones**, **vitamins**, **water**, **salts** and **urea**. Blood cells and large molecules, e.g. plasma proteins, remain in the blood.

Ⓑ **Selective reabsorption** occurs in the nephrons. **Useful substances** are reabsorbed from the filtrate back into the blood travelling through the capillaries wrapped around each nephron:

- **Glucose**, **amino acids**, **hormones**, **vitamins**, **some water** and **some salts** are reabsorbed in the first convoluted tubule.
- **Some water** is reabsorbed in the loop of Henle.
- **Some salts** and **some water** are reabsorbed in the second convoluted tubule.

Water is reabsorbed by **osmosis**, the other substances are reabsorbed by **diffusion** and **active transport**.

Ⓒ **Filtrate** containing **urea**, **excess water** and **excess salts** enters the collecting ducts where some **water** can be reabsorbed from it by osmosis. The filtrate, now called **urine**, travels down the collecting duct to the **ureter** and then to the **bladder** where it is stored.

Figure 10.3 *Detailed structure of a nephron explaining how urine is produced*

Osmoregulation in humans

Osmoregulation is the regulation of the concentration of blood plasma and body fluids. It is essential to prevent water moving into and out of body cells unnecessarily.

- If the body fluids become **too dilute**, water **enters** body cells by osmosis. The cells swell and may burst.
- If the body fluids become **too concentrated**, water **leaves** body cells by osmosis. The cells shrink and the body becomes **dehydrated.** If too much water leaves cells, metabolic reactions cannot take place and cells die.

Table 10.1 *Water gain and loss by the human body*

Water gain	Water loss
• In **drink**.	• From the kidneys in **urine**.
• In **food**.	• From the skin in **sweat**.
• **Metabolic water** is produced by cells during **respiration**.	• From the respiratory system during **exhalation**.

The kidneys and osmoregulation

The **kidneys** regulate the concentration of body fluids by controlling how much **water** is reabsorbed into the blood plasma during **selective reabsorption.** This determines how much water is lost in urine.

- **If body fluids become too concentrated**

  Excessive sweating, drinking too little or eating a lot of salty foods causes body fluids to become too concentrated. The **hypothalamus** of the brain detects that the blood plasma is too concentrated and stimulates the **pituitary gland** at the base of the brain to secrete **antidiuretic hormone (ADH).** The blood carries the ADH to the kidneys where it makes the walls of the second convoluted tubules and the collecting ducts **more permeable** to water. Most of the water is reabsorbed from the filtrate back into the blood and very **small quantities** of **concentrated** urine are produced.

- **If body fluids become too dilute**

  Drinking a lot of liquid makes body fluids too dilute. The hypothalamus detects this and the pituitary gland stops secreting ADH. Without ADH, the walls of the second convoluted tubules and collecting ducts remain almost **impermeable** to water so very little water is reabsorbed from the filtrate back into the blood. **Large quantities** of **dilute** urine are produced.

Regulation of the concentration of blood plasma and body fluids is one aspect of **homeostasis**, i.e. the maintenance of a constant internal environment.

Kidney failure and dialysis

When **kidney failure** occurs, the nephrons stop functioning properly so that they are unable to remove waste from the blood and regulate the volume and composition of blood plasma and body fluids. Harmful waste, especially urea, builds up in the blood and can reach toxic levels resulting in death. Kidney failure can be treated by a **kidney transplant** or **dialysis.**

During **dialysis** blood from a vein, usually in the arm, flows through a **dialysis machine** and is then returned to the body. In the machine, the blood is separated from **dialysis fluid** by a partially permeable membrane. **Waste products**, mainly **urea**, pass from the blood into the dialysis fluid together with **excess water** and **excess salts**. In this way, waste from the blood is removed and the volume and composition of the blood plasma and body fluids are regulated. Dialysis must occur at regular intervals; most people require three sessions a week, each lasting 4 hours.

Revision questions

1. a Define excretion.

   b Explain why egestion is not classified as excretion.

2. Explain what will happen to an organism if excretion does not occur.

3. Explain how plants excrete their metabolic waste.

4. Name the excretory organs in humans.

5. Outline how urine is produced in the kidney.

6. What effect would drinking three large glasses of water have on the quantity and composition of your urine? Explain your answer.

7. A person suffering from kidney failure can be treated using dialysis. Explain why this treatment must occur at regular intervals.

11 Movement

Living organisms must **move** to some extent; some are capable of moving their whole bodies while others only move body parts.

__Movement__ is a change in the position of a whole organism or of parts of an organism.

Types of movement

Living organisms display different types of movement:

- **Growth movement** is a movement displayed by a plant due to the plant growing. **Shoots** grow and bend towards **light** while **roots** grow and bend downwards with the pull of **gravity** (see page 100).
- **Part movement** is a movement displayed by part of a plant or of a sedentary animal. Part movements include the opening and closing of petals, the 'feeding' movements of insectivorous plants, the folding of leaves of the sensitive plant (*Mimosa*) when touched, and the movement of the tentacles of coral polyps and limbs of barnacles.
- **Whole body movement** or **locomotion** is the movement of the entire body from one place to another as occurs in animals. Locomotion is brought about by **muscles** contracting against a **skeleton** in most animals.

The human skeleton

Humans have an **endoskeleton** that lies inside the body, surrounded by muscles. The skeleton is held together at **joints** by tough elastic **ligaments** and is made from **bone** and **cartilage**.

- **Bone** is composed of living cells surrounded by the non-living mineral, calcium hydroxyapatite ($Ca_{10}(PO_4)_6(OH)_2$), together with some collagen (protein) fibres. Blood vessels run throughout bone.
- **Cartilage** is composed of living cells surrounded mainly by collagen fibres, which make it more elastic than bone.

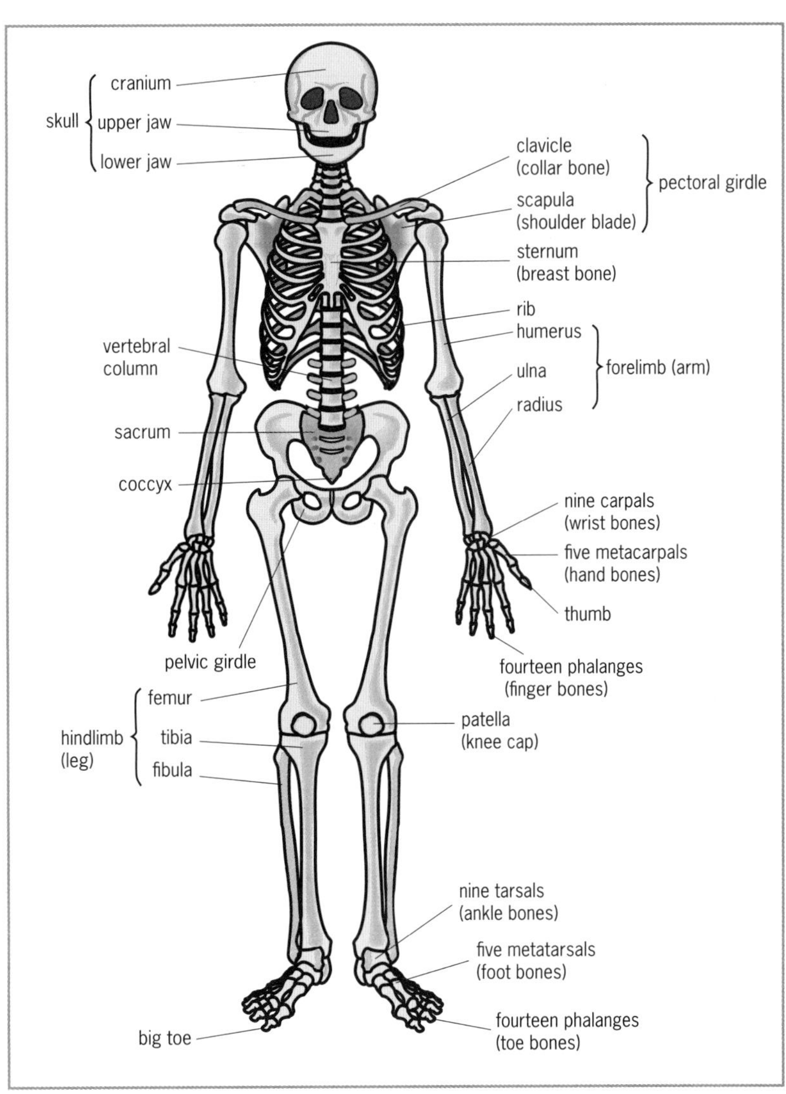

Figure 11.1 *The human skeleton*

The skeleton can be divided into the **axial skeleton** and the **appendicular skeleton**.

The axial skeleton

The axial skeleton consists of the **skull**, **vertebral column**, **ribs** and **sternum**:

- The **skull** is made up of the **cranium** and **upper jaw** which are fused, and the **lower jaw** which articulates with the upper jaw. The skull protects the brain and the sense organs of the head.
- The **vertebral column** is composed of 33 bones known as **vertebrae**. The column supports the body, provides points of attachment for the girdles and many muscles, and protects the spinal cord that runs through it. It also allows some movement.
- The **ribs** are attached to the vertebral column dorsally and the **sternum** ventrally. They form a curved, bony cage that protects the heart and lungs, and movement of the ribs is essential for breathing (see page 68).

The appendicular skeleton

The appendicular skeleton is composed of the **pectoral girdle**, the **pelvic girdle**, the **arms** (forelimbs) and the **legs** (hindlimbs). Both the arms and the legs are built on the same basic pattern known as the **pentadactyl limb**.

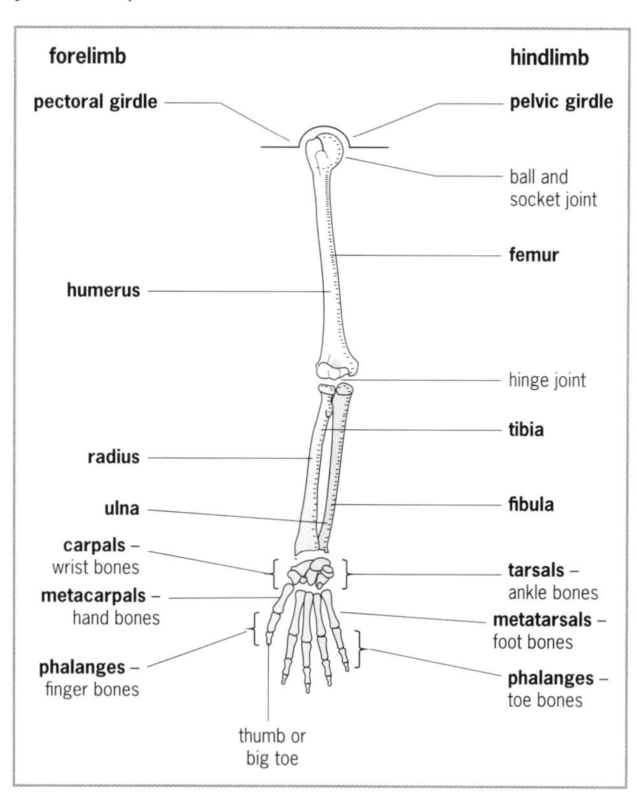

Figure 11.2 *The pentadactyl limb*

- The **girdles** connect the limbs to the axial skeleton. The pelvic girdle is fused to the sacrum at the bottom of the vertebral column to provide **support** for the lower body and to transmit the **thrust** from the legs to the vertebral column which moves the body forwards. Both girdles have broad, flattened surfaces for the **attachment** of muscles that move the limbs.
- The **limbs** are composed of long bones that have **joints** between to allow for easy **movement**. Being long, the bones provide a large surface area for the **attachment** of muscles and permit long **strides** to be taken.

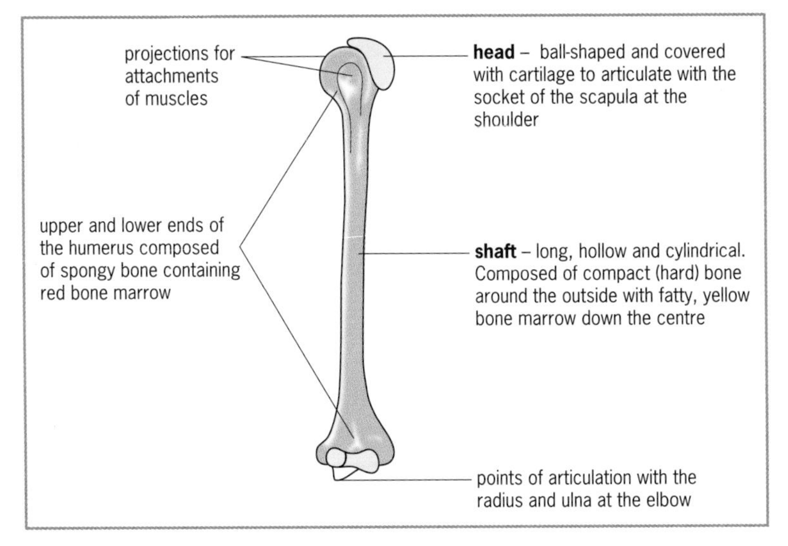

Figure 11.3 *Structure of the humerus*

Functions of the human skeleton

The human skeleton has **four** main functions:

- **Protection** for the internal organs. The skull protects the brain and sense organs of the head, the vertebral column protects the spinal cord, and the ribs and sternum protect the lungs and heart.
- **Support** for the soft parts of the body. This is mainly carried out by the vertebral column, pelvic girdle and legs.
- **Movement**. The skeleton is jointed and muscles work across these joints to bring about movement. The vertebral column and limbs are mainly responsible for movement.
- **Manufacture of blood cells**. Red blood cells, most white blood cells and platelets are manufactured in the red bone marrow found in flat bones, e.g. the pelvis, scapula, ribs, sternum, cranium and vertebrae, and in the ends of long bones, e.g. the humerus and femur.

Movement in humans

Movement in humans is brought about by **skeletal muscles** working across **joints**.

Joints

Joints are formed where two bones meet. There are **three** main types of joints:

- **Immoveable joints** or **fibrous joints**. The bones are joined firmly together by fibres allowing no movement, e.g. the cranium is made of several bones joined by immovable joints.
- **Partially movable joints** or **cartilaginous joints**. The bones are separated by **cartilage pads** which allow slight movement, e.g. the vertebrae are separated by intervertebral discs of cartilage.
- **Moveable joints** or **synovial joints**. The articulating surfaces of the bones are covered with **articular cartilage** and are separated by **synovial fluid**. The bones are held together by **ligaments**, which are tough and **elastic** to prevent dislocation, but still allow movement. There are two types of moveable joints:
    - **Hinge joints** are formed when the ends of bones meet. They allow movement in one plane and are capable of bearing heavy loads, e.g. the elbow, knee, finger and toe joints.
    - **Ball and socket joints** are formed where a ball at the end of one bone fits into a socket in the other bone. They allow movement in all planes, e.g. the shoulder and hip joints.

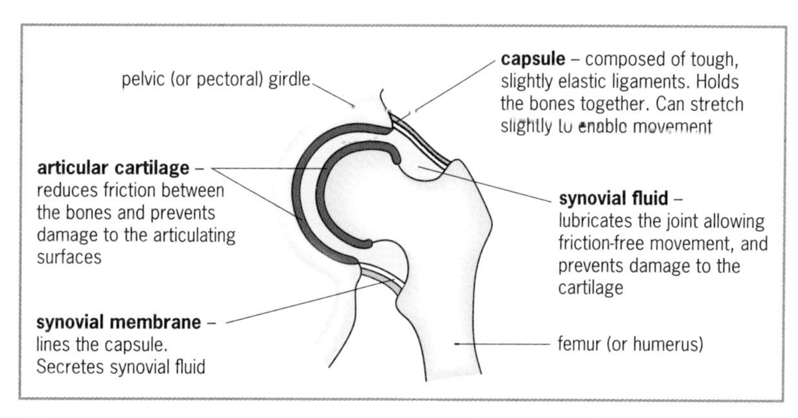

Figure 11.4 *Structure and function of the parts of a ball and socket joint*

Skeletal muscles

Skeletal muscles consist of bundles of multinucleate muscle fibres which are surrounded by connective tissue. **Tendons** attach these muscles to the bones of the skeleton. Tendons are tough and **non-elastic** so that when a muscle contracts the force is transmitted directly to the bone, causing it to move.

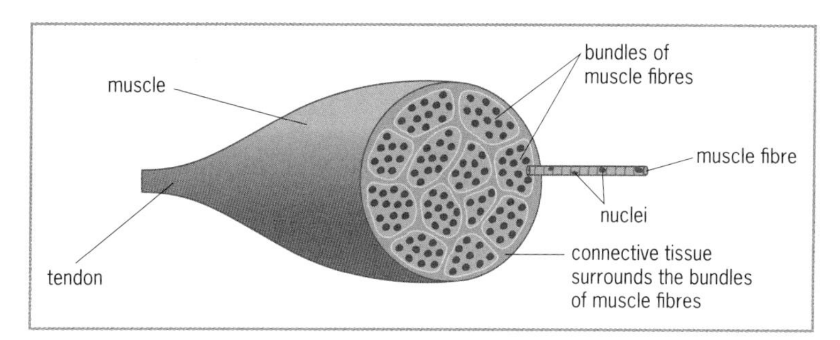

Figure 11.5 *A portion of a skeletal muscle*

Action at moveable joints

When a muscle **contracts** it exerts a **pull**, but it cannot exert a push when it relaxes. Therefore, **two** muscles are always needed to produce movement at a moveable joint, known as an **antagonistic pair**. The muscle that **bends** the joint when it contracts is called the **flexor muscle**. The muscle that **straightens** the joint when it contracts is called the **extensor muscle**.

The **origin** of a muscle is the attachment point of the end of the muscle to a bone that does not move during contraction. The **insertion** is the attachment point of the muscle to the bone that moves.

Movement of the human forelimb

The **biceps** and **triceps** muscles move the radius and ulna causing the elbow joint to bend or straighten.

- The **biceps** is the **flexor** muscle. Its origin is on the scapula, which does not move, and its insertion is on the radius close to the elbow joint.
- The **triceps** is the **extensor** muscle. Its origin is on the scapula and top of the humerus, which do not move, and its insertion is on the ulna close to the elbow joint.

To **bend** the elbow joint, the biceps contracts and the triceps relaxes. To **straighten** the elbow joint, the triceps contracts and the biceps relaxes.

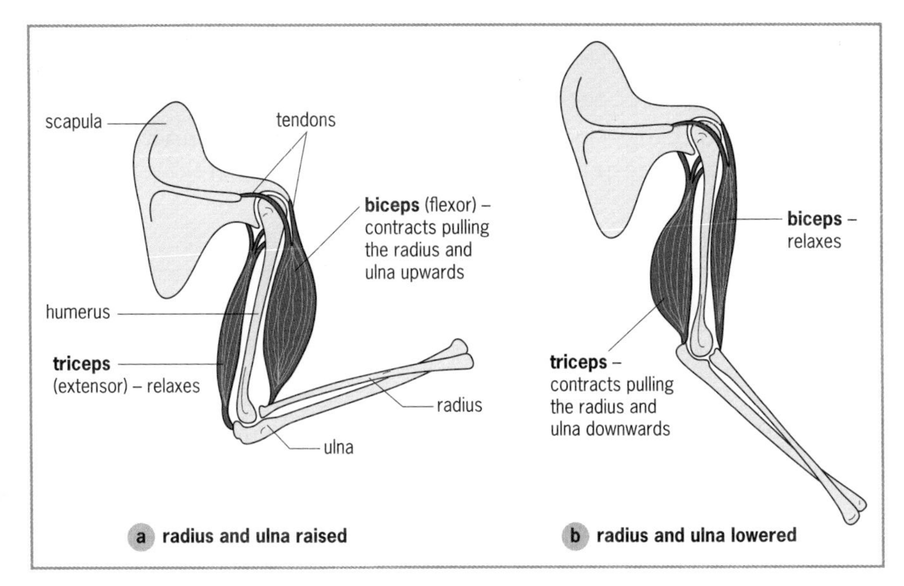

Figure 11.6 *Movement of the lower forelimb*

The importance of locomotion in animals

The ability of animals to move their whole bodies from place to place is important for several reasons:

- Animals move to **search for food**. Plants make their own food by photosynthesis so their roots must remain anchored in the soil to absorb water and mineral salts.
- Animals move to **search for a mate** for sexual reproduction. Plants rely on external agents such as insects and wind to carry their pollen grains for reproduction.
- Animals move to **escape from predators**. Some plants have developed mechanisms to protect against being eaten such as spines or toxic substances that make parts of them poisonous.
- Animals move to **distribute their offspring**. Many plants rely on external agents such as animals, wind and water to disperse their seeds.
- Animals also move to:
  - Prevent overcrowding.
  - Avoid danger.
  - Avoid their waste products.
  - Avoid harsh environmental conditions.
  - Colonise new habitats.

Revision questions

1. Distinguish between growth movements in plants and movement in animals.
2. By referring to the different parts of the human skeleton, discuss its FOUR main functions.
3. What is a joint?
4. Name the THREE different types of joints found in the human skeleton and give ONE location of EACH type.
5. Why is it important that ligaments are slightly elastic, but tendons are non-elastic?
6. Using TWO labelled and annotated diagrams, explain how the muscles in the human arm bring about bending and straightening of the elbow joint.
7. Give FOUR reasons why it is important that animals are able to move their whole bodies from place to place.

12 Irritability

All living organisms must be able to **respond** to changes in their environment to help them survive. Animals can respond quickly to these changes whereas the responses of plants are generally much slower.

Definitions

- ***Stimulus***: *a change in the internal or external environment of an organism that initiates a response.*
- ***Response***: *a change in an organism or part of an organism which is brought about by a stimulus.*
- ***Receptor***: *the part of the organism that* ***detects*** *the stimulus.*

  In animals, the **sense organs** contain the receptors, e.g. the eyes, ears, nose, tongue and skin contain specialised receptor cells that detect stimuli. In plants, the very **tips** of roots and shoots act as receptors.
- ***Effector***: *the part of an organism that* ***responds*** *to the stimulus.*

  In animals, **muscles** and **glands** are effectors. In plants, the regions **just behind** the tips of roots and shoots and the **petioles** of leaves are effectors.

Responses of green plants to stimuli

Plants respond to stimuli by making **part movements** or **growth movements**, which aid in survival.

Part movements

Changes in the **turgidity** of cells bring about many part movements:

- The **leaves** of some plants respond to touch or strong winds by folding, which protects them from damage, e.g. *Mimosa* (the sensitive plant).
- The **leaves** of some plants respond to changing light intensities by folding at night and opening in the morning to access light for photosynthesis, e.g. tamarind.
- The **flowers** of some plants respond to changing light intensities by opening in the morning to expose the stamens and carpels for pollination, and closing at night, e.g. hibiscus. Others open at night and close in the morning, e.g. night flowering cactus.
- Parts of **insectivorous plants** move to trap prey, e.g. Venus fly trap snaps closed to trap insects.

Growth movements

Growth movements are brought about by parts of plants **growing** in response to a stimulus.

- **Shoots** grow and bend **towards** unilateral **light**. This maximises the amount of light they have available for **photosynthesis**. In the absence of light, or when illuminated evenly, shoots grow **upwards** against **gravity** to 'search' for light or to maximise the amount of light they receive. Growing upwards also ensures that flowers are held in the best position for pollination, and fruits and seeds for seed dispersal.
- **Roots** grow and bend **downwards** with the pull of **gravity**. This enables them to **anchor** the plant in the ground and to obtain **water** and **minerals**. Roots also grow and bend **towards water** to maximise the amount of water they can obtain for **photosynthesis**.

Responses of invertebrates

Invertebrates such as millipedes, earthworms and woodlice move their whole bodies towards or away from stimuli. These responses aid the survival of the organisms and can be investigated using a **choice chamber** in which the organisms are provided with adjacent environments with different environmental conditions, e.g. dry or moist. The organisms are placed in the centre of the chamber and their distribution is recorded after a fixed length of time.

Table 12.1 *Responses of invertebrates*

Stimulus	Response	How the response aids survival
Light	Most move away from the light into darkness.	Makes the organisms harder to be seen by predators.
Moisture	Most move away from dry areas into areas with moisture.	Prevents desiccation (drying out), especially if the organisms do not have waterproof body coverings.
Temperature	Move away from very low or very high temperatures.	Prevents extremes of temperature affecting enzyme activity.
Chemicals	Move towards chemicals given off by food and away from harmful chemicals.	Enables organisms to find food which is essential for survival, and avoid being harmed by chemicals such as pesticides.
Touch	Move away or curl up if touched.	Helps the organisms escape from predators or gives protection against predators.

The nervous system of humans

The human nervous system is composed of **neurones** or **nerve cells** and is divided into two parts:

- The **central nervous system (CNS)** which consists of the **brain** and the **spinal cord**.
- The **peripheral nervous system (PNS)** which consists of **cranial** and **spinal nerves** that connect the central nervous system to all parts of the body.

Neurones

Neurones make up both the CNS and PNS, and they transmit messages called **nerve impulses**. All neurones have a **cell body** with thin fibres of cytoplasm extending from it called **nerve fibres**. Nerve fibres that carry impulses **towards** the cell body are called **dendrites**. Nerve fibres that carry impulses **from** the cell body are called **axons**; each neurone has only one axon. There are **three** types of neurones:

- **Sensory neurones** which transmit impulses from **receptors** to the **CNS**.
- **Motor neurones** which transmit impulses from the **CNS** to **effectors**.
- **Relay** or **intermediate neurones** which transmit impulses throughout the **CNS**. They link sensory and motor neurones.

Nerves are made up of bundles of nerve fibres of sensory and/or motor neurones surrounded by connective tissue. The **brain** and **spinal cord** are made up mainly of relay neurones and the cell bodies of motor neurones.

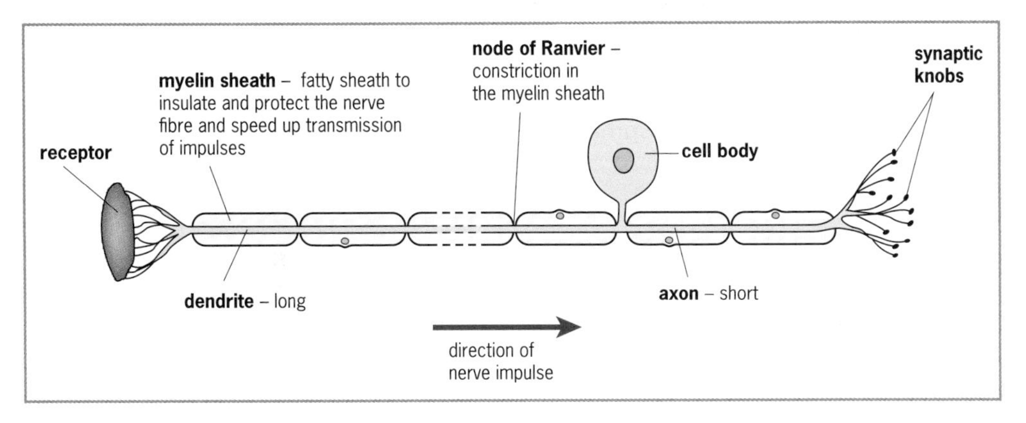

Figure 12.1 *Structure of a sensory neurone*

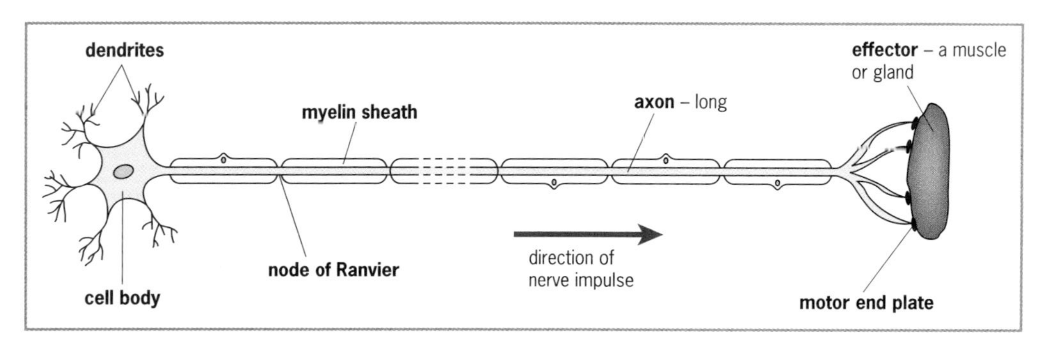

Figure 12.2 *Structure of a motor neurone*

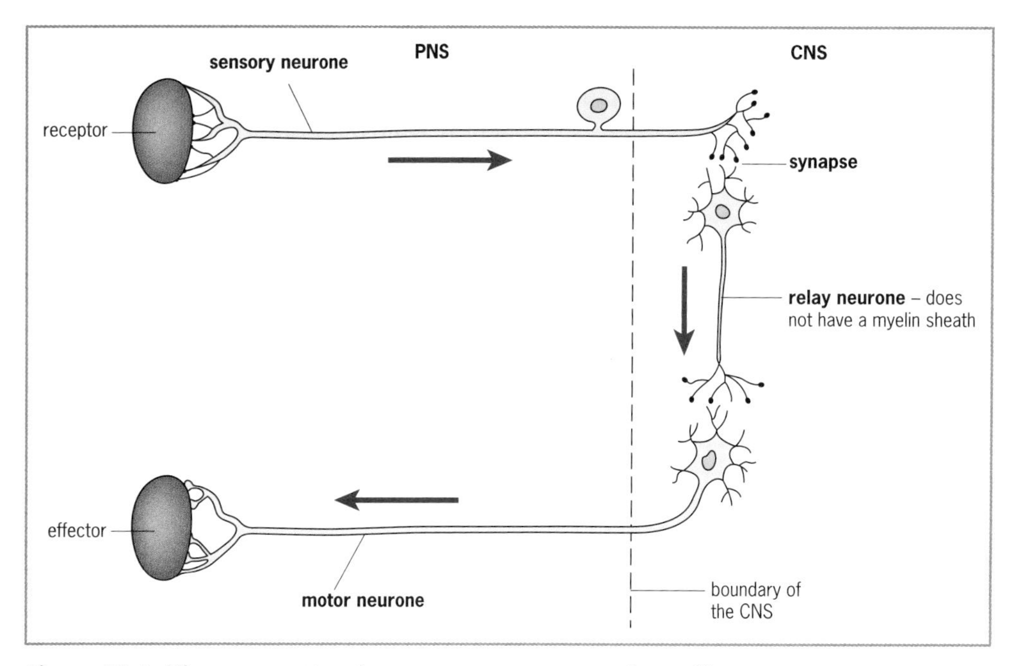

Figure 12.3 *The connection between a receptor and an effector*

Transmitting impulses between neurones

Adjacent neurones do not touch. There are tiny gaps called **synapses** between the synaptic knobs at the end of one axon and the dendrites or cell body of adjacent neurones. **Chemicals** are released into the synapses by the synaptic knobs. These chemicals cause impulses to be set up in adjacent neurones. This ensures impulses travel in **one direction** only.

Coordinating function of the central nervous system

The job of the central nervous system is to **coordinate** the activities of all parts of the body. It gathers information from receptors via sensory neurones. It then **processes** this information and sends messages out to effectors via motor neurones so that the most appropriate action can be taken. Messages are passed between the brain and the spinal cord by relay neurones.

Simple reflex actions

A **reflex action** is a rapid, automatic, involuntary response to a stimulus by a muscle or gland, e.g. the automatic withdrawal of the hand when it touches a hot object. Simple reflex actions happen without conscious thought, they are not learned and they aid in survival. The pathway between receptor and effector is known as a **reflex arc** and it involves the following:

- A **receptor** which detects the stimulus.
- A **sensory neurone** which carries the impulse to the central nervous system.

- A **relay neurone** in the central nervous system, which carries the impulse to a motor neurone.
- A **motor neurone** which carries the impulse away from the central nervous system.
- An **effector** which responds to the stimulus.

Simple reflexes are classified as **cranial reflexes** or **spinal reflexes**.

Cranial reflexes

In cranial reflexes, impulses pass through the **brain**, e.g. the pupil reflex, blinking, sneezing, coughing and saliva production.

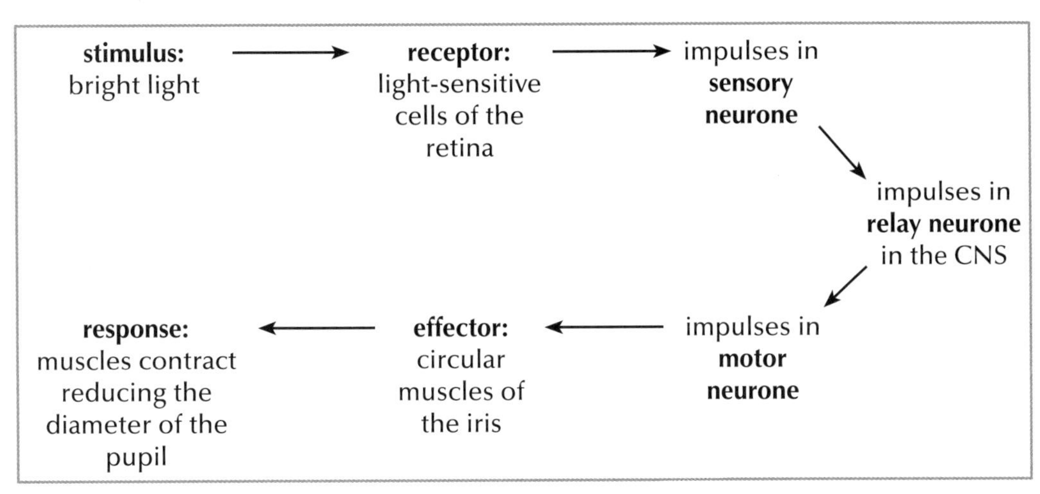

Figure 12.4 *A simple flow diagram to illustrate the pupil reflex*

Spinal reflexes

In spinal reflexes, impulses pass through the **spinal cord**, e.g. the knee jerk reflex and the withdrawal reflex.

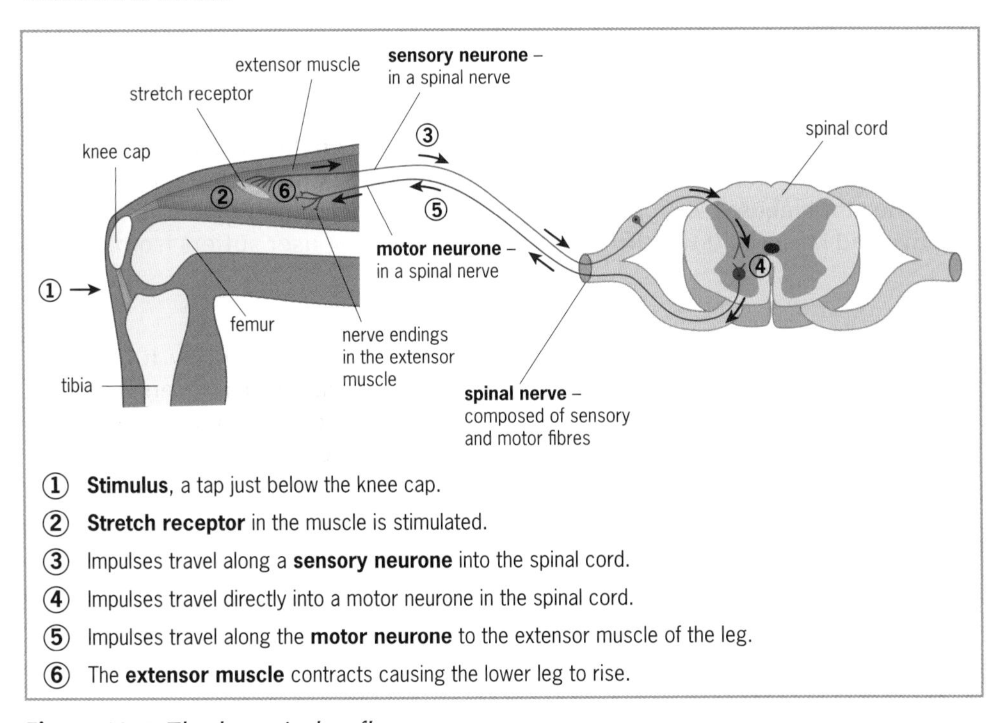

① **Stimulus**, a tap just below the knee cap.
② **Stretch receptor** in the muscle is stimulated.
③ Impulses travel along a **sensory neurone** into the spinal cord.
④ Impulses travel directly into a motor neurone in the spinal cord.
⑤ Impulses travel along the **motor neurone** to the extensor muscle of the leg.
⑥ The **extensor muscle** contracts causing the lower leg to rise.

Figure 12.5 *The knee jerk reflex*

The human brain

The human brain is an extremely complex organ with different regions concerned with different functions. The main regions are:

- the **cerebrum** which is composed of two **cerebral hemispheres**
- the **cerebellum**
- the **medulla**.

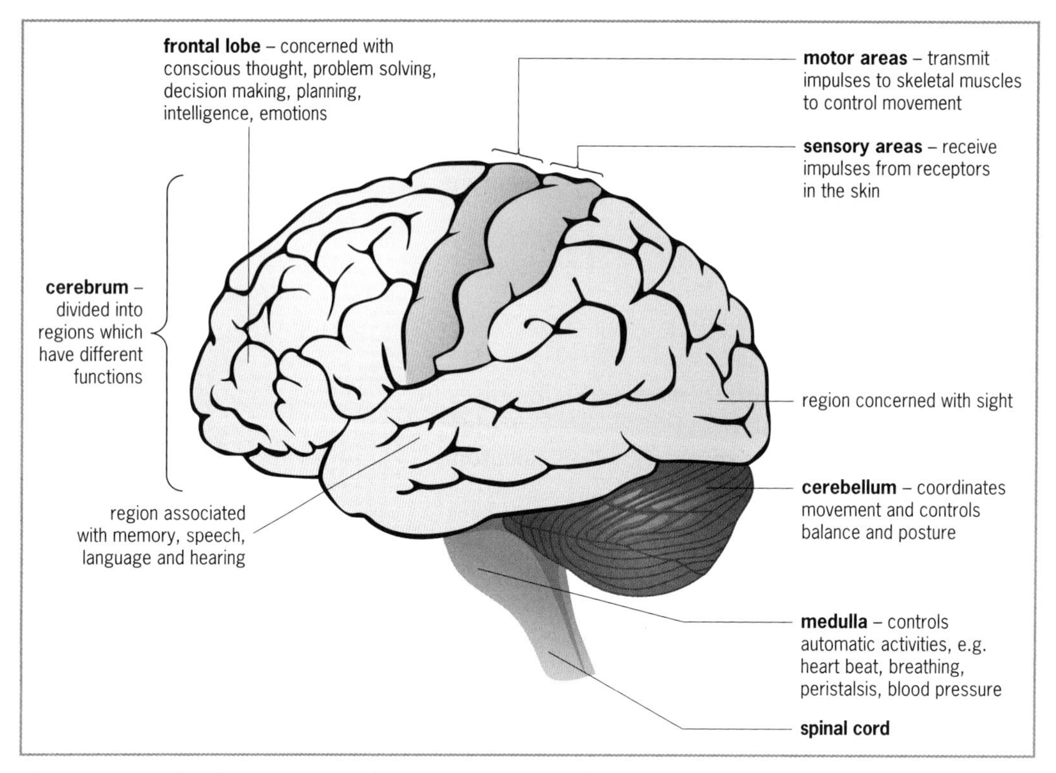

Figure 12.6 *The functions of the main regions of the human brain*

Physiological effects of drug abuse

A **drug** is any chemical substance that affects the functioning of the body. Many drugs are used medically to improve health. If **abused**, i.e. used wrongly, all drugs, including prescription drugs, can become addictive and harmful. If the drug is then withdrawn, the user suffers from **withdrawal symptoms**.

Abuse of alcohol

Alcohol is a **depressant** of the central nervous system. A person who repeatedly consumes alcoholic beverages in excess of normal social drinking customs runs the risk of becoming an **alcoholic**. Alcoholics are dependent on alcohol and **alcoholism** is classed as a **disease**. Severe alcoholism can lead to death. Alcohol abuse can cause the following:

Short-term effects

- Impaired muscular skills, reduced muscular coordination and slowed reflexes.
- Impaired mental functioning, concentration and judgement.
- Blurred vision and slurred speech.
- Memory lapses.
- Drowsiness.

- Increased urine production leading to dehydration.
- Loss of consciousness.

Long-term effects

- Long-term memory loss.
- Increased blood pressure causing heart disease, heart attack and stroke.
- Inflammation of the stomach walls, stomach ulcers and other intestinal disorders.
- Fatty liver disease, alcoholic hepatitis and cirrhosis (scarring) of the liver.
- Nervous system disorders and brain damage as brain cells die.
- Cancer of the mouth, throat and oesophagus.
- Delirium tremens (DTs), a condition characterised by body tremors, anxiety and hallucinations.

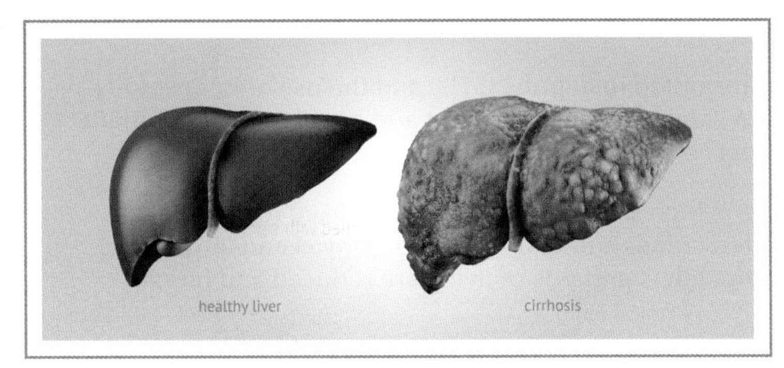

A healthy liver (left) and a liver with cirrhosis (right)

Children born to mothers who are alcoholics may suffer from **foetal alcohol syndrome** which is characterised by mental retardation.

Abuse of cocaine

Cocaine is a highly addictive illegal drug which is a **stimulant** of the central nervous system. Abuse of cocaine can cause the following:

- Feelings of well-being, increased energy, alertness, confidence and power.
- Bizarre, erratic, violent behaviour and hallucinations.
- Paranoia, anxiety and depression.
- Increased breathing rate and heart rate.
- Reduced need for sleep.
- Damage to the lungs and nasal passages.
- Loss of appetite which can lead to nutritional deficiencies and increased susceptibility to infection.
- Constriction of blood vessels which increases blood pressure and can lead to a heart attack or stroke.
- Schizophrenia and other mental disorders.

Abuse of prescription drugs

Prescription drugs that are most commonly abused include tranquillisers, antibiotics, diet pills, pain killers (analgesics), caffeine and steroids.

- **Tranquilisers** are used to treat anxiety and insomnia. Abuse of tranquillisers, e.g. valium, can lead to slurred speech, poor muscular coordination, reduced attention span, dizziness, sleepiness, blurred vision, hallucinations, mental confusion, low blood pressure and apathy.

- **Antibiotics** are used to treat bacterial infections. Overuse or taking an incomplete course of antibiotics can lead to the development of **antibiotic resistant** strains of bacteria. In cases of infection by these resistant strains, antibiotics are of no use. Some people also show allergic reactions to antibiotics, e.g. penicillin, which can be fatal.
- **Diet pills** are used to help lose weight. Abuse of diet pills can lead to nervousness, dizziness, anxiety, difficulty sleeping, diarrhoea, stomach pain, high blood pressure, a fast or irregular heartbeat, heart palpitations and heart failure.

Social and economic effects of drug abuse

Drug abuse upsets relationships with family and friends, and leads to personal neglect, automobile accidents, job loss as the abuser is unable to work, financial problems, increased crime or even prostitution as the abuser has to find money to pay for the drugs, increased demands on health services, and a shortened life span of the abuser.

Prostitution exposes abusers to sexually transmitted infections (STIs) and the use of intravenous drugs exposes them to AIDS and hepatitis A. Crime can lead to arrest and possible imprisonment. Babies born to abusers may have birth defects or be addicted to the drug themselves.

The cost to society of drug abuse in general is high as resources have to be used to treat and rehabilitate drug addicts, to fight drug related crimes and to apprehend, convict and imprison traffickers and pushers of illegal drugs. Ultimately, standards of living are reduced and human resources are lost.

Revision questions

1. Define the terms 'stimulus' and 'response'.
2. Using specific examples where appropriate, explain how BOTH part movements and growth movements aid the survival of plants.
3. Identify THREE stimuli to which invertebrates respond and state how the responses aid survival.
4. Name the THREE types of neurones found in the nervous system and indicate the function of EACH.
5. What is a simple reflex action?
6. By means of a simple flow diagram, show the pathway along which impulses would travel when a student pricks her finger and immediately withdraws her hand from the source of the pain.
7. Name the THREE main regions of the human brain and outline the functions of EACH region.
8. Identify FOUR effects of the abuse of cocaine on the human body.
9. Name FOUR prescription drugs that are often abused.

The human eye

The **eye** detects **light** that has been reflected from an object and converts it into **nerve impulses**. The impulses are transmitted along the **optic nerve** to the brain which translates them into a precise picture of the object.

The eyes are situated in bony sockets of the skull called **orbits** and have muscles attached to move them. The orbits protect the back of each eye from damage, and the **eyelids** and **eyelashes** protect the front from foreign particles. **Tears**, produced by tear glands above each eye, keep the eyes moist, wash away foreign particles and contain an enzyme that destroys micro-organisms.

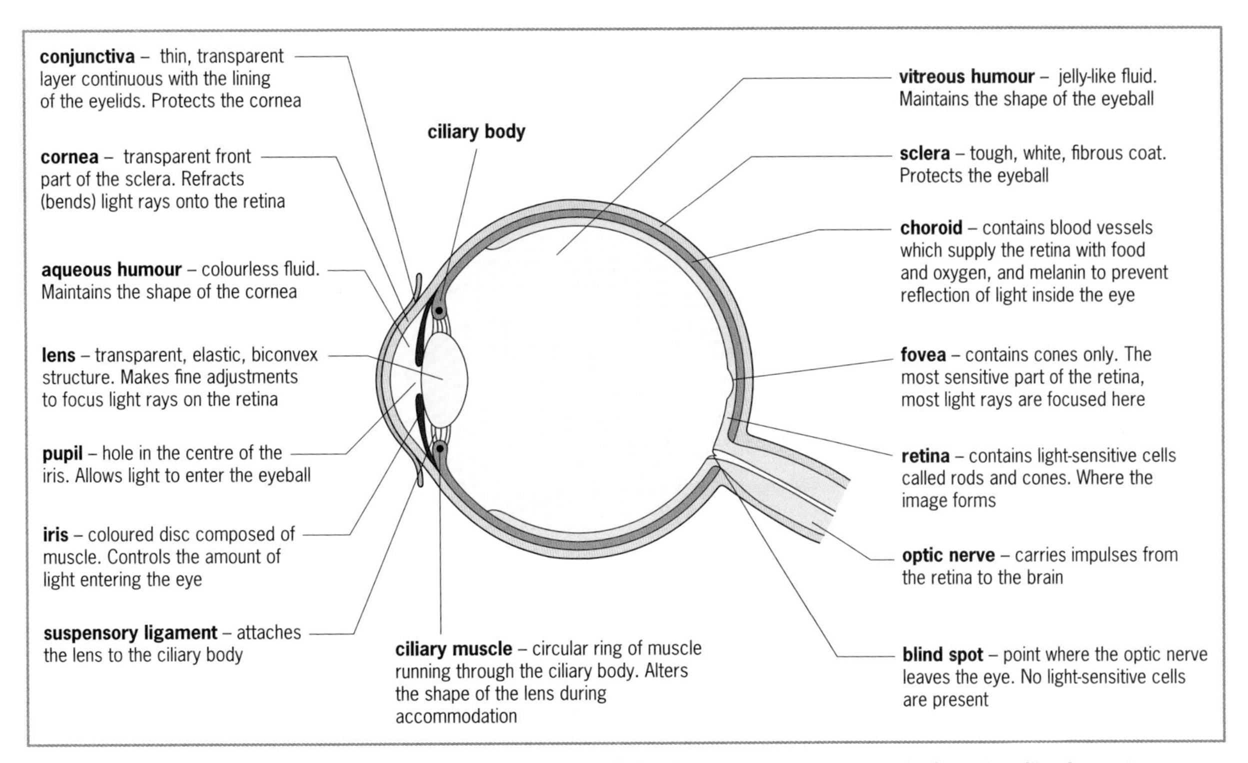

Figure 12.7 *Structure and functions of the parts of the human eye as seen in longitudinal section*

Image formation

In order to see, light rays from an object must be **refracted** (bent) as they enter the eye so that they form a clear **image** of the object on the receptor cells of the retina. Being **convex** in shape, both the **cornea** and the **lens** refract the light rays.

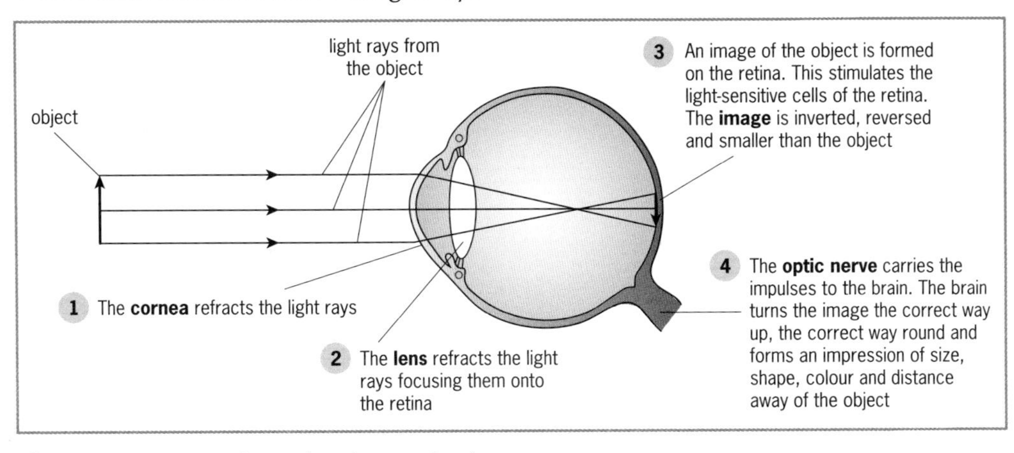

Figure 12.8 *Formation of an image in the eye*

Detection of light intensity and colour by the eye

The **retina** is composed of two types of specialised **light-sensitive cells** or **photoreceptors**:

- **Rods** function in **low light intensities**. They are responsible for detecting the **brightness** of light and are located around the sides of the retina. Images falling on the rods are seen in shades of black and white only.
- **Cones** function in **high light intensities**. They are responsible for detecting **colour** and **fine detail**, and are mainly located around the back of the retina. The **fovea** is composed entirely of cones which are packed closely together. There are three types that detect either the red, green or blue wavelengths of light.

Focusing light onto the retina – accommodation

By changing shape, the **lens** makes fine adjustments to focus the light rays onto the retina. Changing the shape of the lens to focus light coming from different distances onto the retina is called **accommodation** and it is brought about by the **ciliary muscles**.

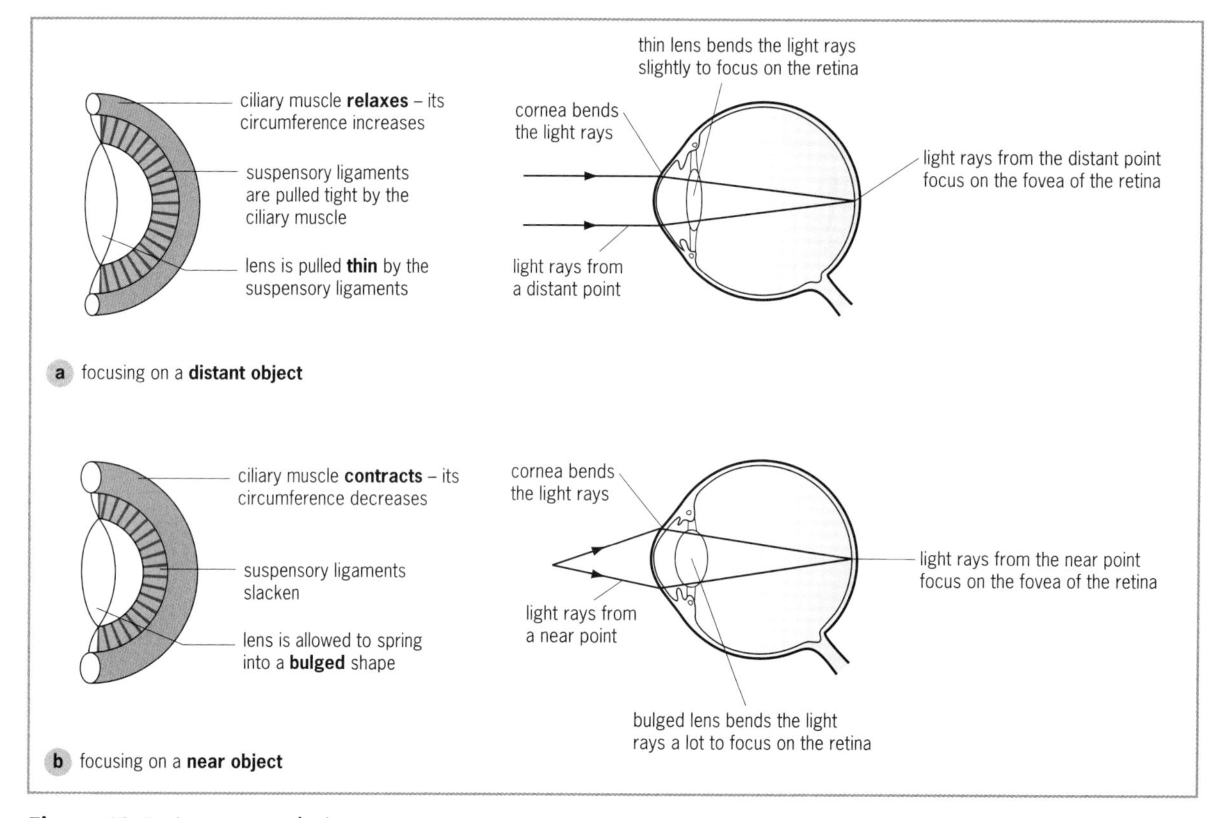

Figure 12.9 *Accommodation*

Control of light entering the eye

The size of the **pupil** controls the amount of light entering the eye. Muscles of the **iris** control the pupil size.

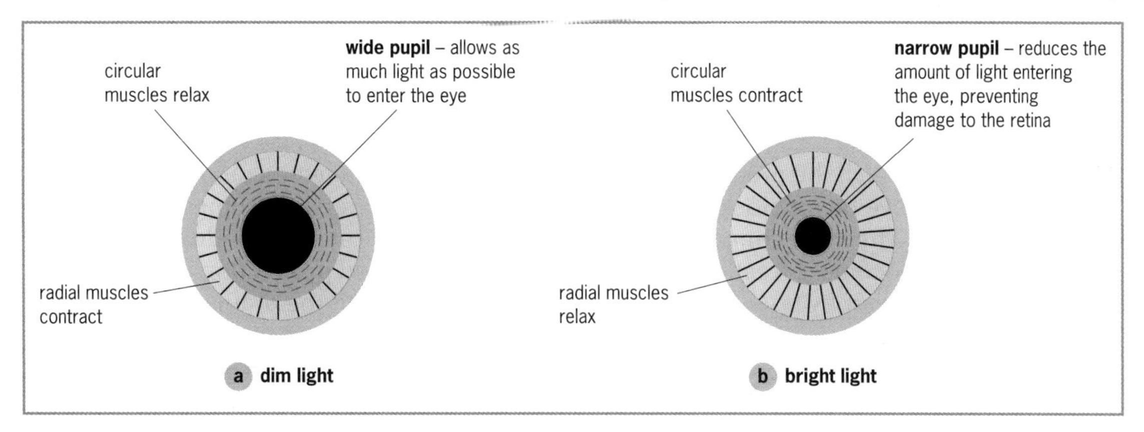

Figure 12.10 *Controlling the amount of light entering the eye*

Sight defects and how they are corrected

Short sight (myopia)

A person with short sight can see **near** objects, but distant objects are out of focus. Light rays from near objects focus on the retina; light rays from **distant** objects focus **in front** of the retina. Short sight occurs if the eyeball is too **long** from front to back or if the lens is too **curved** (thick). It is corrected by wearing **diverging (concave) lenses** as spectacles or contact lenses.

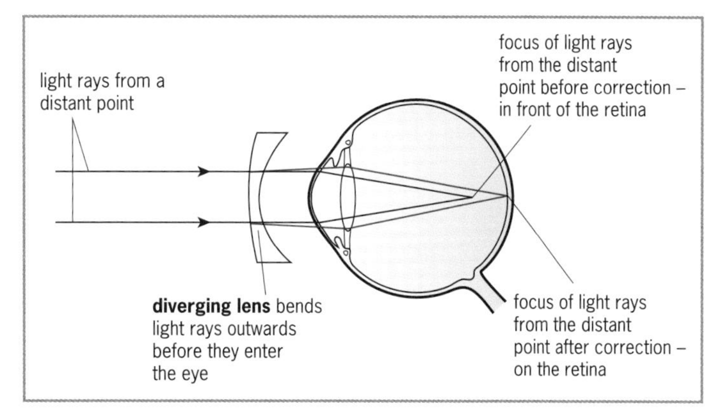

Figure 12.11 *The cause and correction of short sight*

Long sight (hypermetropia)

A person with long sight can see **distant** objects, but near objects are out of focus. Light rays from distant objects focus on the retina; light rays from **near** objects focus **behind** the retina. Long sight occurs if the eyeball is too **short** from front to back or if the lens is too **flat** (thin). It is corrected by wearing **converging (convex) lenses** as spectacles or contact lenses.

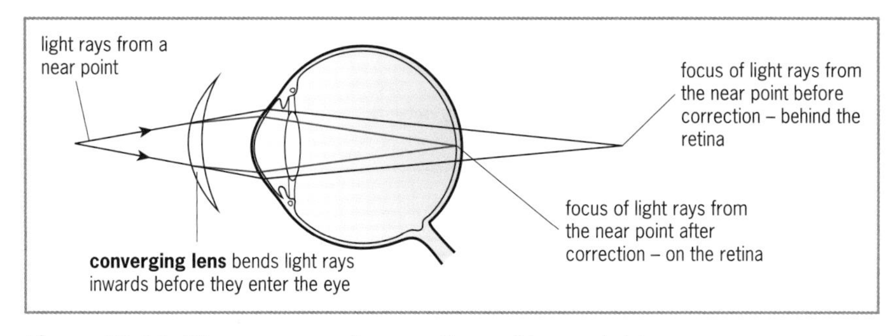

Figure 12.12 *The cause and correction of long sight*

Old sight (presbyopia)

As a person **ages**, the lens loses its elasticity and the ciliary muscles weaken. The lens is less able to curve and the person finds it increasingly difficult to see near objects. It is corrected by wearing **converging lenses** for looking at near objects.

Glaucoma

Glaucoma is a condition in which the **pressure** of the fluid within the eye increases due to the flow of aqueous humour from the eye being blocked. If left untreated, the optic nerve becomes damaged and it can lead to **blindness**. The most common type develops slowly and causes a gradual loss of peripheral (side) vision. Glaucoma is treated with eye drops to reduce fluid production or improve the flow of fluid from the eye, or by laser treatment or an operation to open the drainage channels.

The human skin

The **skin** is the largest organ in the human body. It is made up of **three** layers:

- the **epidermis** which is the outermost layer
- the **dermis** which is below the epidermis
- the **subcutaneous layer** which is the bottom layer made up mainly of fat cells.

The skin plays a vital role in detecting various stimuli such as touch, protecting the body against infection by pathogens and excessive water loss, and regulating body temperature.

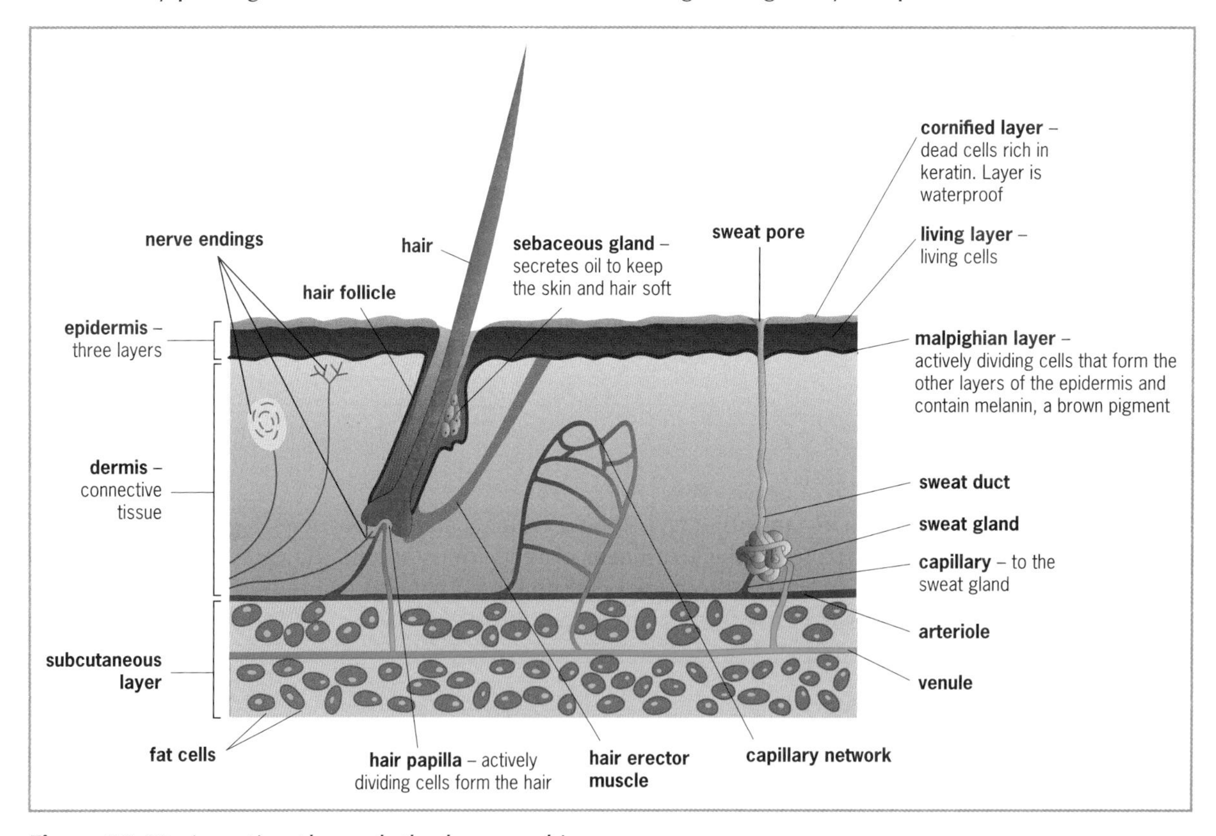

Figure 12.13 *A section through the human skin*

The skin and temperature regulation

Living organisms must maintain a **constant** internal body temperature for **enzymes** to function correctly. In humans this temperature is about 37 °C. Humans, as well as other mammals and birds, are **homeotherms**, i.e. they can maintain a fairly constant body temperature regardless of the environmental temperature. They **gain** most of their body heat from internal **metabolic processes**, mainly respiration, and the blood carries this heat around the body. Heat is **lost** from the body mainly by conduction, convection and radiation through the skin, and evaporation of water during breathing and sweating.

The human skin plays a major role in regulating the body temperature. The **hypothalamus** of the brain detects if the body temperature rises above 37 °C or drops below 37 °C. It sends messages along nerves to the skin causing the responses summarised in Table 12.2. The control of body temperature is one aspect of **homeostasis**, i.e. the maintenance of a constant internal environment.

Table 12.2 *How the skin helps to maintain a constant body temperature*

Body temperature rises above 37 °C	Body temperature drops below 37 °C
Sweating occurs: secretory cells in the sweat glands absorb water, urea and salts from the blood passing through them forming sweat. Sweat passes up the sweat ducts and onto the surface of the skin where the water **evaporates** and removes heat from the body.	**Sweating stops**: there is no water to evaporate and remove heat from the body.
Vasodilation occurs: arterioles and capillaries in the dermis of the skin **dilate** so more blood flows through them and more heat is lost to the environment from the blood.	**Vasoconstriction occurs**: arterioles and capillaries in the dermis of the skin **constrict** so very little blood flows through them and very little heat is lost. The heat is retained by the blood flowing through vessels deeper in the body.
Hair erector muscles relax: this causes the hairs to **lie flat** so no insulating layer of air is created.	**Hair erector muscles contract**: this causes the hairs to **stand up** and trap a layer of air next to the skin that acts as insulation. This is important in hairy mammals and creates 'goose bumps' in humans.

In addition to the responses in Table 12.2, the following occur:

- If the body temperature **rises** above 37 °C, the **metabolic rate decreases** so that less heat is produced in the body.
- If the body temperature **drops** below 37 °C, the **metabolic rate increases** so that more heat is produced and **shivering** may occur in muscles to generate even more heat.

The skin and protection

The skin plays a major role in **protecting** the body:

- The epidermis protects against the entry of **pathogens**.
- The epidermis, being waterproof, protects against **water loss** by evaporation from body fluids.
- The **melanin** in the epidermis protects against the sun's harmful **ultra-violet rays**.
- The epidermis protects against **harmful chemicals** in the environment.
- The subcutaneous layer protects against **heat loss** in low environmental temperatures.
- The subcutaneous layer protects against **damage** by acting as 'padding'.

Skin care

Good **skin care** can help delay the natural ageing process and prevent skin problems.

- **Clean** the skin daily:
  - Use **warm water**. Limit exposure to hot water which removes natural oils.
  - Use **mild cleansers**. Avoid strong soaps which remove natural oils.
  - **Pat** the skin dry so that some moisture remains.
- **Moisturise** the skin daily with a moisturiser suitable for the skin type. Moisturisers contain ingredients such as **humectants** which absorb moisture from the environment, **oils** and **lubricants**. These are designed to make the epidermis softer and more pliable by increasing its water content.
- Apply **sunscreen** daily to protect the skin from the sun's harmful ultra-violet rays. Use a sunscreen with a minimum **sun protection factor (SPF)** of 15. SPF is a laboratory measure of the effectiveness of the sunscreen, the higher the SPF the more effective the sunscreen should be.
- Avoid skin care products containing **alcohol** which has a drying effect.
- Use skin care products and sunscreens that contain **natural ingredients**. Avoid products containing **synthetic chemical ingredients** thought to be harmful to the body, e.g. parabens, polyethylene glycols (PEGs) and siloxanes found in moisturisers, and oxybenzone found in sunscreens.

Skin bleaching

Skin bleaching refers to the practice of using chemical substances to **lighten skin tone** or to provide an **even skin complexion** by reducing the production of melanin in areas of abnormal pigmentation, e.g. in moles, birthmarks and other darker patches. The chemicals usually work by inhibiting the action of an **enzyme** (tyrosinase) that is necessary for the formation of melanin.

The safety of some chemicals used is questionable. Many bleaching creams and lotions contain **hydroquinone** which has been found to cause **mutations** and **cancer** in animals. Hydroquinone can also **irritate** the skin, cause **skin sensitivity** and cause blue-black skin **discolouration**. Some bleaching products contain **steroids** which can cause the skin to become **thin**.

Revision questions

10 By means of a fully labelled and annotated diagram only, indicate the functions of the different parts of the human eye.

11 Explain how the eye enables us to see.

12 How do the eyes of a person adjust when he:

a walks from a dimly lit room into the bright sunshine
b looks at a book in his hand after watching an aeroplane in the sky?

13 Explain how short sight is caused and how it is corrected.

14 What is glaucoma?

15 Describe the changes occurring in the skin if the body temperature rises above 37 °C.

16 Give FOUR ways in which the skin protects the body.

17 Outline how you should take care of your skin on a daily basis.

18 What is skin bleaching?

13 Growth

All living organisms **grow** and **develop**.

__Growth__ is a permanent increase in the size of an organism.

To grow, the cells of multicellular organisms undergo **cell division** by **mitosis** (see page 145). The new cells then grow to full size by manufacturing more **protoplasm**. Plant cells increase in size even more by absorbing water into their vacuoles. Most cells then **differentiate (specialise)** to carry out specific functions. Animals also grow by making more **extracellular materials**, e.g. mineral of bones, fibres of connective tissue.

Methods of measuring growth

Methods used to measure growth involve measuring changes in various growth parameters over time.

Table 13.1 *Methods used to measure growth*

Parameter being measured	Used to measure growth of	Advantages	Disadvantages
Height	Humans. Small plants.	Quick to measure. Easy to measure. Doesn't harm the organism.	Only measures growth in one dimension. Difficult to determine the top of a plant.
Length	Bodies of animals. Parts of animals, e.g. legs, wings, tails. Stems. Leaves.	Quick to measure. Easy to measure. Doesn't harm the organism.	Only measures growth in one dimension. Only measures the growth of part of an organism.
Wet mass	Most animals. Small plants.	Quick to measure. Easy to measure. Doesn't harm animals. Gives a more accurate measure of overall growth than height or length.	Measurements may be inconsistent due to changes in water content of the bodies of organisms. Plant growth is disturbed when plants are uprooted and the roots cleaned.
Dry mass	Germinating seeds and seedlings. Small plants. Small animals, e.g. invertebrates.	Gives the most accurate measure of growth because it measures cellular and extracellular material without water.	Time consuming; organisms have to be dried at 100 °C to constant mass. Organisms are killed. Large numbers of organisms are required.
Number of leaves	Small plants.	Relatively quick. Easy to count.	Only measures one aspect of growth.

Parameter being measured	Used to measure growth of	Advantages	Disadvantages
Surface area of leaves	Plants.	Gives a more accurate measure of overall growth than leaf length.	Time consuming; leaves have to be outlined on squared paper and the squares counted.
Number of organisms	Populations.	Relatively quick.	Can be difficult for animal populations since animals move.

Individual organs, organisms and populations display similar growth patterns. If the parameter used to measure growth is plotted against time, a **growth curve** is obtained which is described as being **sigmoid** or **S-shaped**.

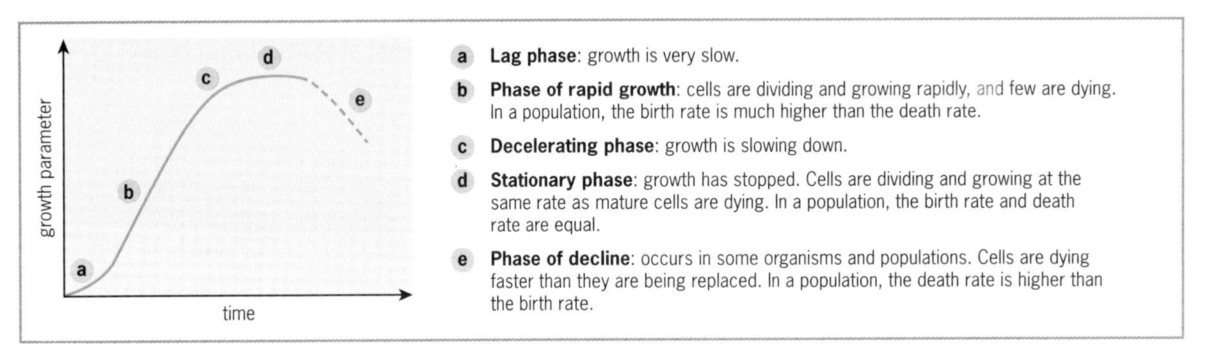

Figure 13.1 *A sigmoid growth curve*

Growth in plants

Growth in plants occurs in two distinct phases:

- The initial period of growth that occurs after fertilisation and forms the **embryo** within the seed (see page 131). The seed may then remain **dormant** for a long period.
- The growth that occurs from the time the seed begins to **germinate**.

Germination

Germination is the process by which the embryonic plant in a seed grows into a seedling.

A **seed** contains the **embryo** of the plant. In dicotyledonous seeds, the embryo consists of a **radicle** or embryonic root and a **plumule** or embryonic shoot which are joined to two **cotyledons**. The cotyledons contain **stored food** to be used by the radicle and plumule during germination. The stored food may include protein, starch and/or lipid. The radicle, plumule and cotyledons are surrounded by the **testa** or seed coat.

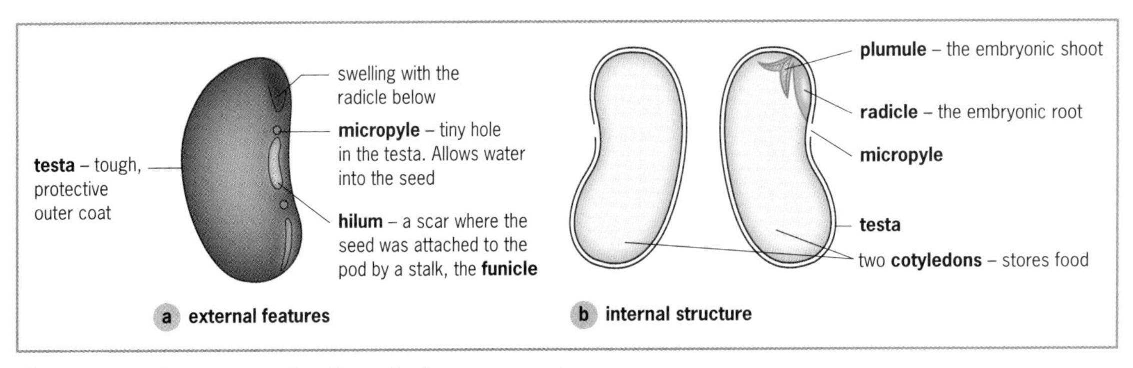

Figure 13.2 *Structure of a dicotyledonous seed*

Seeds require **three conditions** to germinate:

- **Water** to activate the enzymes so that chemical reactions can occur.
- **Oxygen** for aerobic respiration to produce energy.
- A **suitable temperature**, usually between about 5 °C and 40 °C, to activate enzymes.

Water is absorbed through the **micropyle** which causes the seed to swell and it activates enzymes that breakdown stored food in the cotyledons. Proteins are broken down into **amino acids**, starch is broken down into maltose and then **glucose**, and lipids into **fatty acids** and **glycerol**. These soluble substances are then **translocated** to the radicle and plumule (see page 87).

- The **amino acids** are used in making new cells in the tips of the radicle and plumule so that **growth** can occur.
- The **glucose** is used in **respiration** to produce **energy** for the radicle and plumule to grow, and to make the **cellulose cell walls** of new cells.
- The **fatty acids** and **glycerol** are used in **respiration**.

As the **radicle** grows, it emerges from the testa and grows downwards. The **plumule** then emerges and grows upwards. The cotyledons may be carried up and out of the soil or they may remain below the soil.

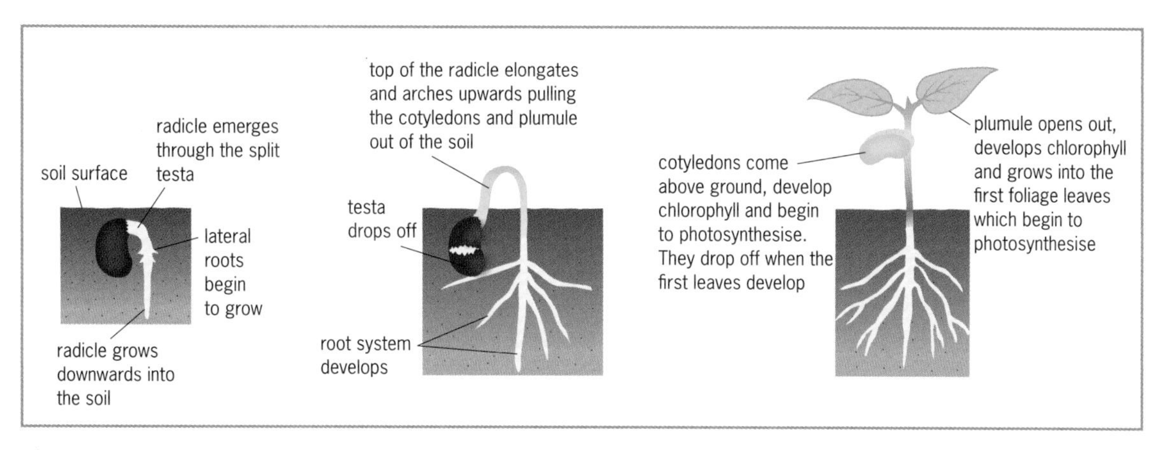

Figure 13.3 *Stages in the germination of a green bean*

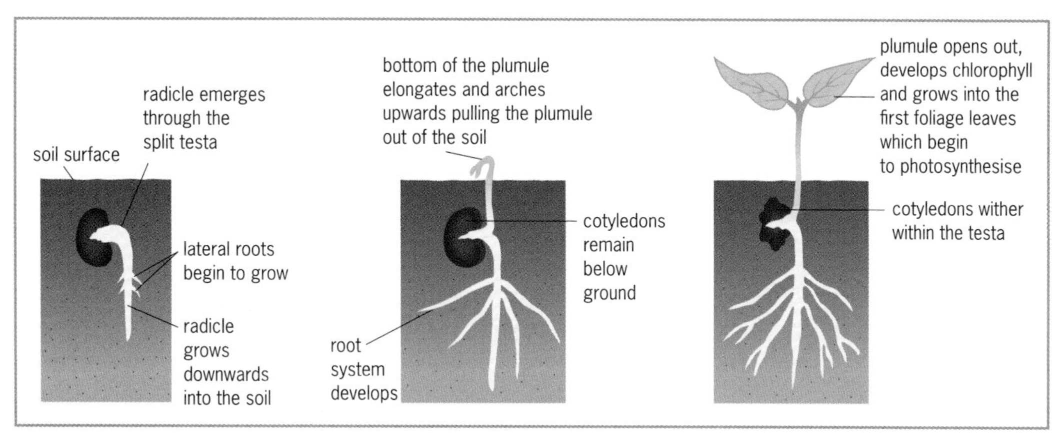

Figure 13.4 *Stages in the germination of a pigeon pea*

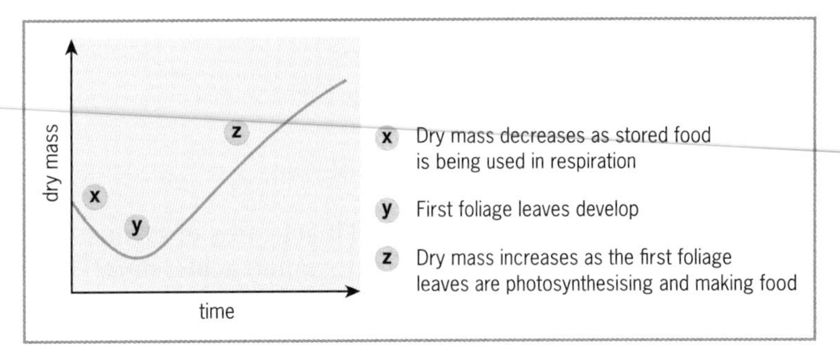

Figure 13.5 *Growth curve of a germinating seed*

Meristems

Plant cells have semi-rigid cell walls that restrict their ability to divide and grow. Consequently, plants have groups of **immature cells** which have thin walls and that retain the ability to actively divide and grow. These groups of cells, known as **meristems**, are found in specific locations where they form the only actively growing tissues in plants.

Growth in length of roots and shoots – primary growth

Growth in length of roots and shoots occurs at their tips. Cells in regions called **apical meristems** in the very tips constantly divide by **mitosis** (see page 145). Newly formed cells then rapidly **elongate** in the region directly behind the apical meristems. They do this mainly by absorbing water into their vacuoles. Once fully elongated, the cells then **differentiate** into xylem and phloem.

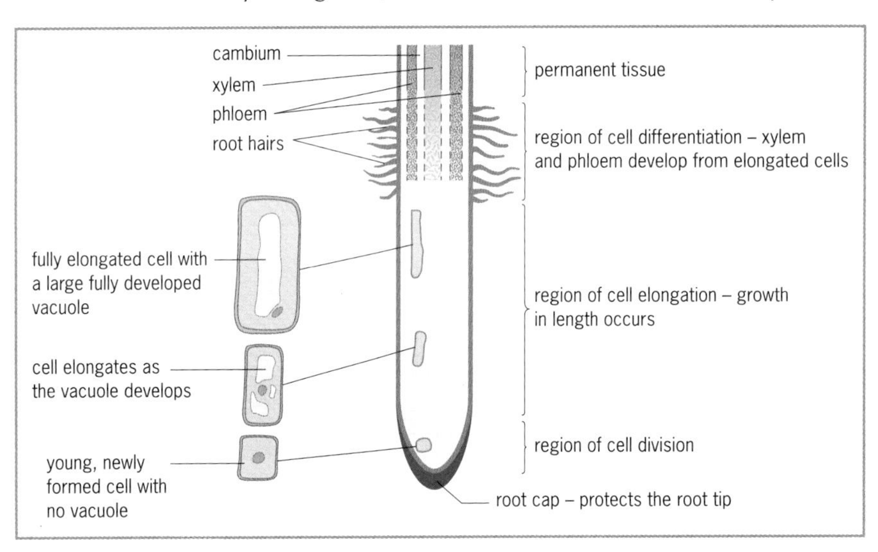

Figure 13.6 *Growth in length of a root*

Growth in width of roots and stems – secondary growth

Roots and stems of plants living for a number of years, such as trees, grow in **width**. Cells in **two** regions called **lateral meristems** constantly divide by mitosis:

- Cells of the **vascular cambium**, found between the xylem and phloem, divide and differentiate into xylem tissue towards the inside and phloem tissue towards the outside. In stems, xylem is added in rings; each ring represents a season's growth (see Figure 9.9, page 83).
- Cells of the **cork cambium**, found below the bark, divide and produce cork or bark to the outside.

A comparison of growth in plants and animals

There are several differences in the way plants and animals grow.

Table 13.2 *Plant and animal growth compared*

Feature	Plants	Animals
Duration of growth	Usually grow continuously throughout their lifetime.	Usually grow to a maximum size and stop growing.
Where cell division occurs	Occurs only in **meristems** which are found in the **tips** of roots and shoots and in the **vascular cambium** and **cork cambium** of roots and stems.	Occurs in most tissues throughout the body.
How growth occurs	Occurs mainly by cells taking water into their vacuoles and expanding.	Occurs mainly by cells increasing in number.

Revision questions

1. What is meant by the term 'growth'?
2. Wet mass and dry mass are two ways growth can be measured. Indicate the advantages and disadvantages of EACH of these methods.
3. What is the name given to the shape of a typical growth curve?
4. Describe the structure of a dicotyledonous seed.
5. Identify the THREE conditions that seeds need in order to germinate and explain the role of EACH condition in the germination process.
6. Describe the internal and external changes that occur when a named seed germinates until the first foliage leaves are formed above ground.
7. What are meristems?
8. Explain how roots grow in length.
9. Give THREE ways in which growth in animals differs from growth in flowering plants.

14 Reproduction

Living organisms must **produce offspring** in order for their species to survive.

***Reproduction** is the process by which living organisms generate new individuals of the same kind as themselves.*

There are two types of reproduction:

- **asexual reproduction**
- **sexual reproduction.**

Asexual and sexual reproduction compared

Asexual reproduction

Asexual reproduction involves only **one** parent and offspring are produced by **mitosis** (see page 145). All offspring produced asexually from one parent are **genetically identical** and are collectively called a **clone**. Asexual reproduction is **conservative** because it conserves the characteristics of the parent. Certain plants (see page 147) as well as fungi and unicellular organisms, e.g. amoeba and bacteria, reproduce asexually.

Sexual reproduction

Sexual reproduction involves **two** parents. **Gametes**, or sex cells, are produced in reproductive organs by **meiosis** (see page 147). A male and a female gamete fuse during **fertilisation** to form a single cell called a **zygote**. The zygote divides by **mitosis** to form an **embryo** and ultimately an **adult**. Offspring produced sexually receive genes from both parents, therefore they possess characteristics of both parents, i.e. they show **variation**.

Table 14.1 *Asexual and sexual reproduction compared*

Asexual reproduction	Sexual reproduction
Produces **no variation** among offspring; all are identical. Consequently: • If the parent is well adapted to its environment, **all** offspring will be well adapted and the chances of them **all** surviving will be high. **(A)** • If environmental conditions change adversely, **all** offspring will be adversely affected, reducing the chances of survival of **all** offspring. **(D)** • It does not enable species to change and adapt to changing environmental conditions. **(D)**	Produces **variation** among offspring; no two organisms are identical. Consequently: • Some offspring may be better adapted to their environment than their parents, others may not be as well adapted, consequently they do **not all** have an equal chance of survival. **(D)** • If environmental conditions change, **some** offspring may be better adapted to the new conditions, increasing the chances of survival of **some** offspring. **(A)** • It enables species to **change** and adapt to changing environmental conditions. **(A)**
The process is **rapid**. It does not involve finding a mate, producing gametes, fertilisation and development of an embryo. This results in a rapid increase in numbers of organisms in populations. (**A**)	The process is **slow**. A mate has to be found in animals, gametes have to be produced, fertilisation has to occur and the embryo has to develop. This results in a slow increase in numbers of organisms in populations. (**D**)
Offspring usually remain **close** to the parent, which increases the chances of overcrowding and competition. (**D**)	Offspring are usually **dispersed** over a wide area, which reduces the chances of overcrowding and competition. (**A**)

(**A**) = advantage (**D**) = disadvantage

Many plants use **both** methods of reproduction, thus combining the **advantages** of both, e.g. many species of grass reproduce asexually by producing runners and sexually by producing flowers and seeds.

- **Asexual reproduction** does not allow species to change and adapt to changing conditions. It is beneficial in a stable, unchanging environment, or in an under-crowded environment.
- **Sexual reproduction** enables species to change and adapt to changing conditions. It is beneficial in an unstable, changing environment, or in an overcrowded environment.

Sexual reproduction in humans

The female reproductive system

The female gametes are called **ova** and they are produced in two **ovaries** which form part of the female reproductive system.

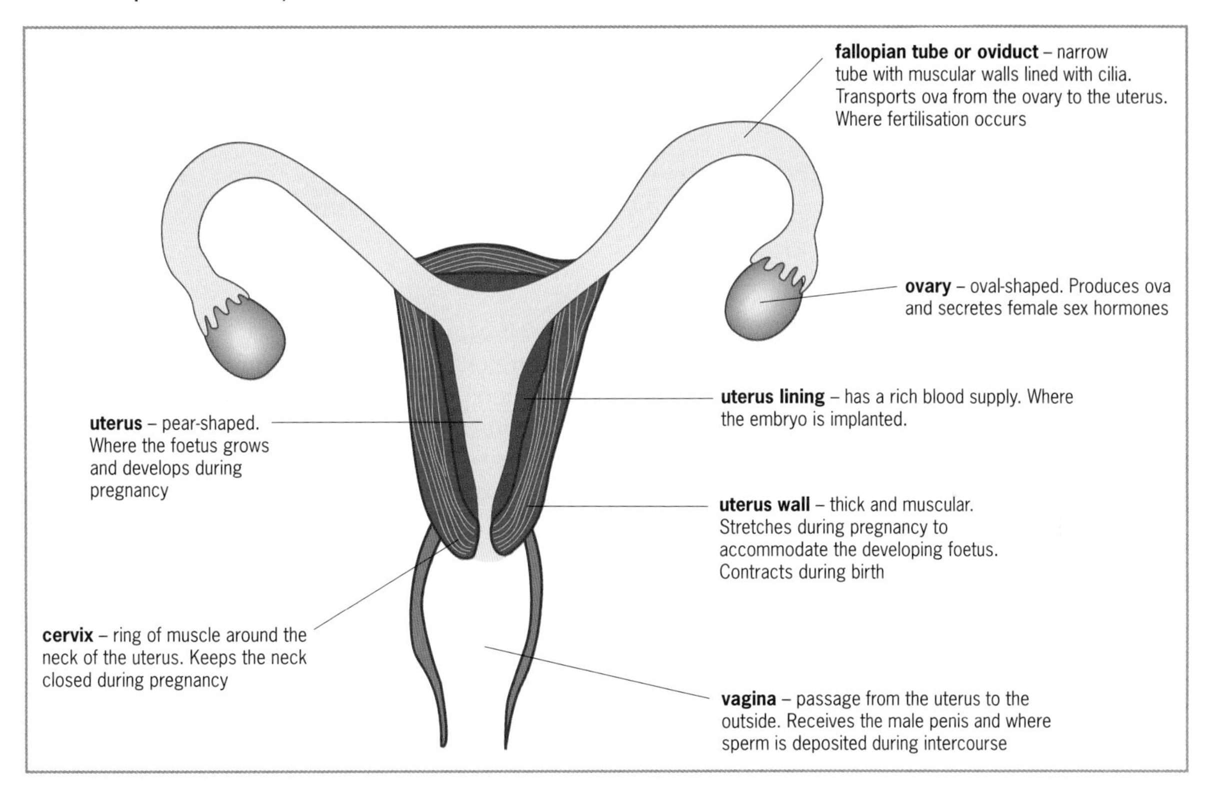

Figure 14.1 *Structure and function of the parts of the female reproductive system*

The male reproductive system

The male gametes are called **sperm** or **spermatozoa** and they are produced in the **testes** which form part of the male reproductive system. Unlike the ovaries which are inside the female body, the testes are located outside the body in a sac called the **scrotum**. This keeps the sperm at a slightly lower temperature than body temperature which is essential for their proper development.

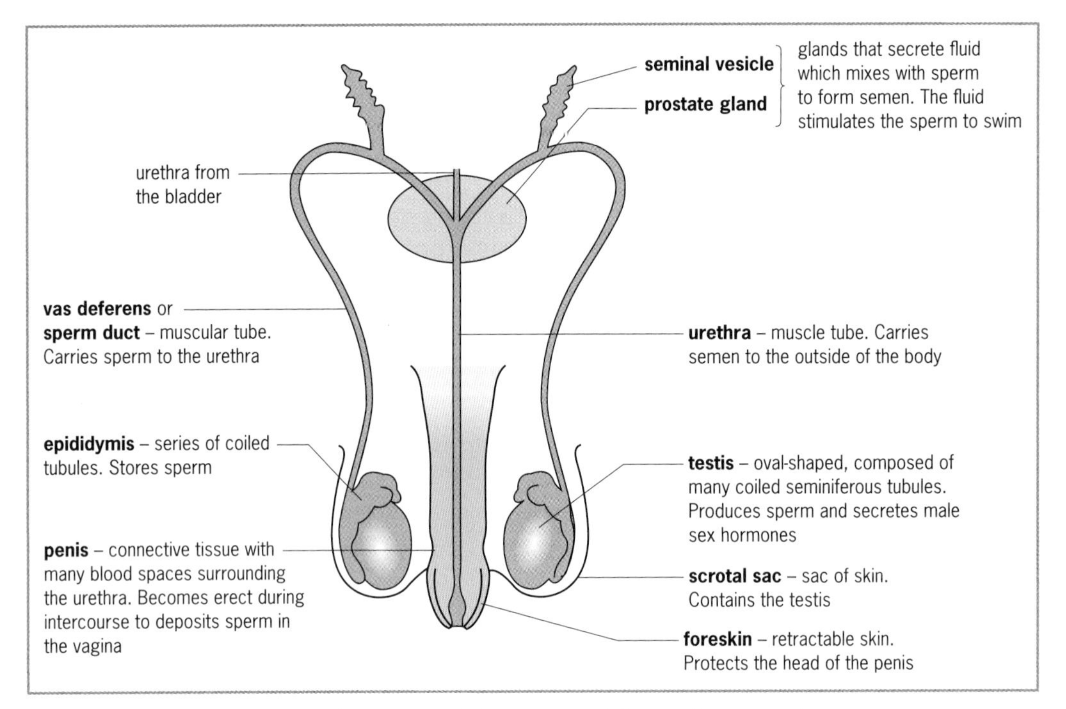

Figure 14.2 *Structure and function of the parts of the male reproductive system*

Production of ova

At birth, each female ovary contains many thousand **immature ova.** Each is surrounded by a fluid-filled space which forms a **primary follicle**. Each month between **puberty** at about 11 to 13 years old, and **menopause** at about 45 to 50 years old, one immature ovum will develop into a **mature ovum.** About 450 immature ova will ever mature. To produce a mature ovum, the immature ovum undergoes **meiosis** (see page 147). One of the four cells produced develops into a mature ovum which is released during **ovulation.**

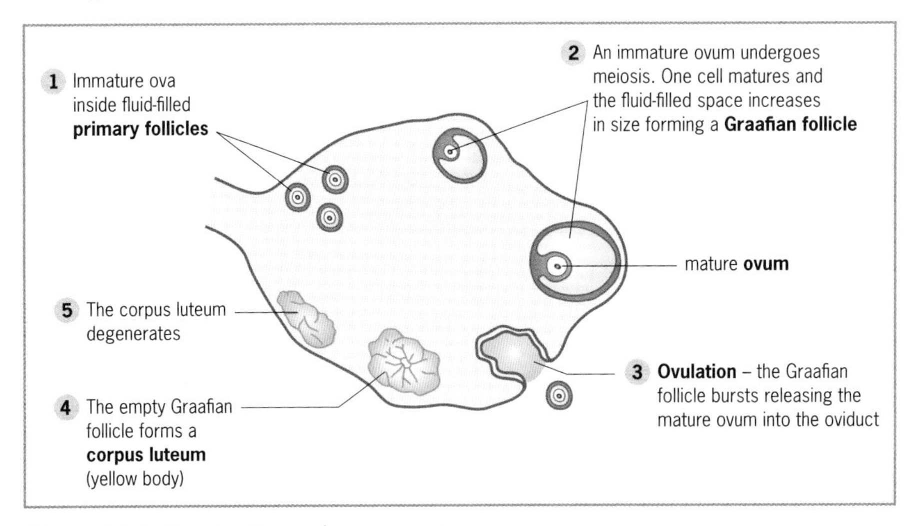

Figure 14.3 *Production of an ovum in an ovary*

Production of sperm

Sperm cells are produced continuously from **puberty** in the **seminiferous tubules** of the **testes**. Cells in the tubule walls undergo **meiosis** and **all** the cells produced develop into mature sperm. These are stored in the **epididymis** until ejaculation.

Table 14.2 *Ova and sperm cells compared*

	Ova	Sperm cells
Structure	layer of follicle cells – act as protection membrane layer of jelly nucleus cytoplasm containing yolk	middle piece – contains mitochondria to release energy for swimming tail – for swimming nucleus small amount of cytoplasm head **acrosome** – contains enzymes to dissolve a passage into the ovum
Production	One is produced each month from puberty to menopause.	Thousands are produced continuously from puberty.
Movement	Are moved down an oviduct after ovulation by muscular contractions of the oviduct walls and beating of the cilia.	Swim actively using their tails when mixed with secretions from the seminal vesicles and prostate gland during ejaculation.
Life span	Live for about 24 hours after ovulation.	Can live for about 2 to 3 days in the female body after ejaculation.

The menstrual cycle

This is a cycle of about 28 days comprising two main events:

- **Ovulation** which is the release of an **ovum** from an ovary.
- **Menstruation** which is the loss of the **uterus lining** from the body. This starts to occur about 14 days after ovulation if fertilisation has not occurred.

The cycle is controlled by **four hormones** which synchronise the production of an ovum with the uterus lining being ready to receive it if fertilised. The start of the cycle is taken from the start of menstruation.

- **Follicle stimulating hormone (FSH)** is secreted by the pituitary gland at the beginning of the cycle.
  - FSH stimulates a Graafian follicle to develop in an ovary and an ovum to mature inside the follicle.
  - FSH stimulates the Graafian follicle to produce **oestrogen.**
- **Oestrogen** is produced by the Graafian follicle mainly during the second week of the cycle.
  - Oestrogen stimulates the uterus lining to thicken and its blood supply to increase after menstruation.
  - Oestrogen causes the pituitary gland to stop secreting FSH and to secrete **luteinising hormone (LH).**
- **Luteinising hormone (LH)** is secreted by the pituitary gland in the middle of the cycle.
  - A sudden rise in LH causes ovulation to take place.
  - LH stimulates the corpus luteum to develop in the ovary and secrete **progesterone**.
- **Progesterone** is produced by the corpus luteum during the third week of the cycle.
  - Progesterone causes the uterus lining to increase slightly in thickness and remain thick.
  - If fertilisation does not occur, the corpus luteum degenerates during the fourth week and reduces secretion of progesterone. The decrease in progesterone causes the uterus lining to begin to break down, and the pituitary gland to secrete **FSH** at the end of the fourth week.

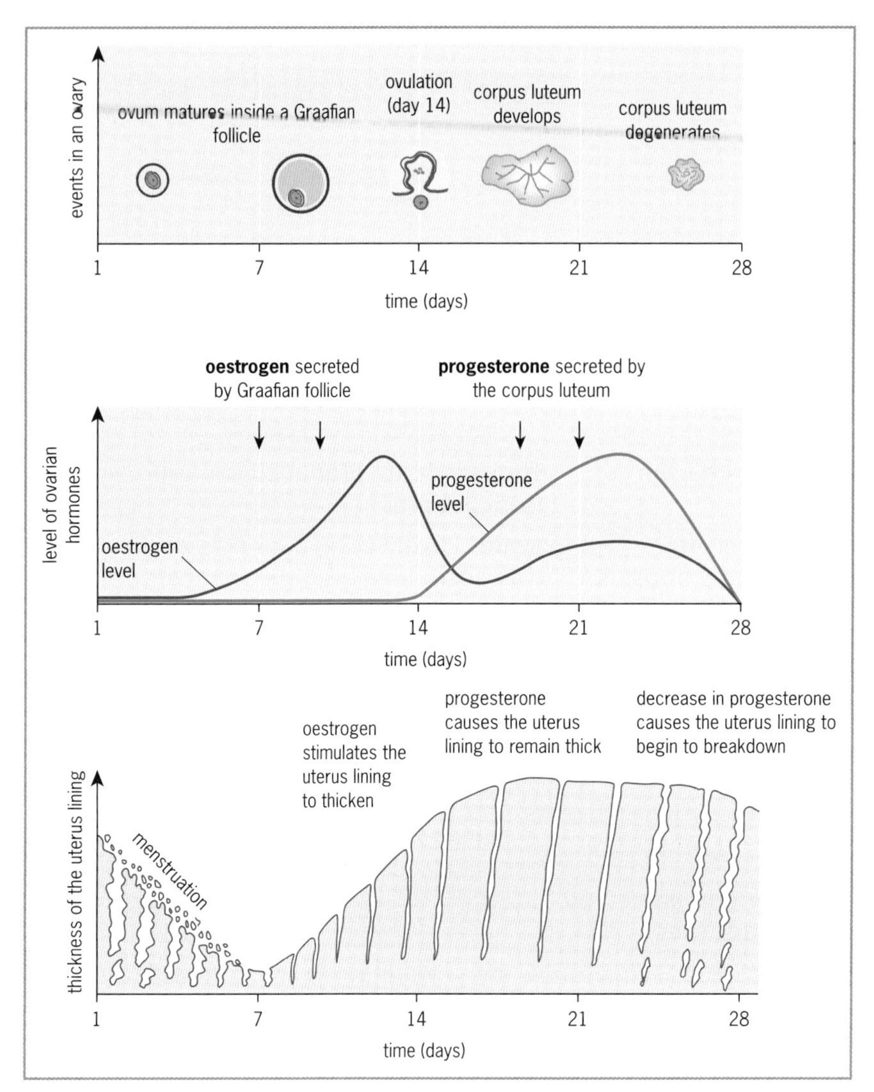

Figure 14.4 *A summary of the events occurring during the menstrual cycle*

Bringing sperm and ova together

When a male becomes sexually excited, blood spaces in the penis fill with blood. The penis becomes **erect** and is placed into the female vagina. **Semen**, composed of sperm and secretions from the seminal vesicles and prostate gland, is **ejaculated** into the top of the vagina by muscular contractions of the tubules of the epididymis and sperm ducts. The **sperm** swim through the cervix and uterus and into the oviducts.

From fertilisation to birth

Fertilisation

If an **ovum** is present in one of the oviducts, one **sperm** enters leaving its tail outside. A **fertilisation membrane** immediately develops around the ovum to prevent other sperm from entering and the nuclei of the ovum and sperm fuse to form a **zygote**.

Implantation

The zygote divides repeatedly by **mitosis** using **yolk** stored in the original ovum as a source of nourishment. This forms a ball of cells called the **embryo** which moves down the oviduct and sinks into the uterus lining, a process called **implantation**. Food and oxygen diffuse from the mother's blood into the embryo and carbon dioxide and waste diffuse back into the mother's blood.

Pregnancy and development

The cells of the embryo continue to divide and some of the cells develop into the **placenta**. The placenta is a disc of tissue with capillaries running throughout and finger-like projections called **villi** that project into the uterus lining. The embryo is joined to the placenta by the **umbilical cord** which has an **umbilical artery** and **umbilical vein** running through. These connect the capillaries in the embryo with those in the placenta.

The **placenta** allows exchange of materials between the mother's blood and the embryo's blood, but prevents mixing of the two bloods which may be of different types. It also prevents certain unwanted substances entering the embryo's blood from the mother's blood, e.g. many bacteria and viruses.

The developing embryo is surrounded by a thin, tough membrane called the **amnion** which forms a sac containing **amniotic fluid**.

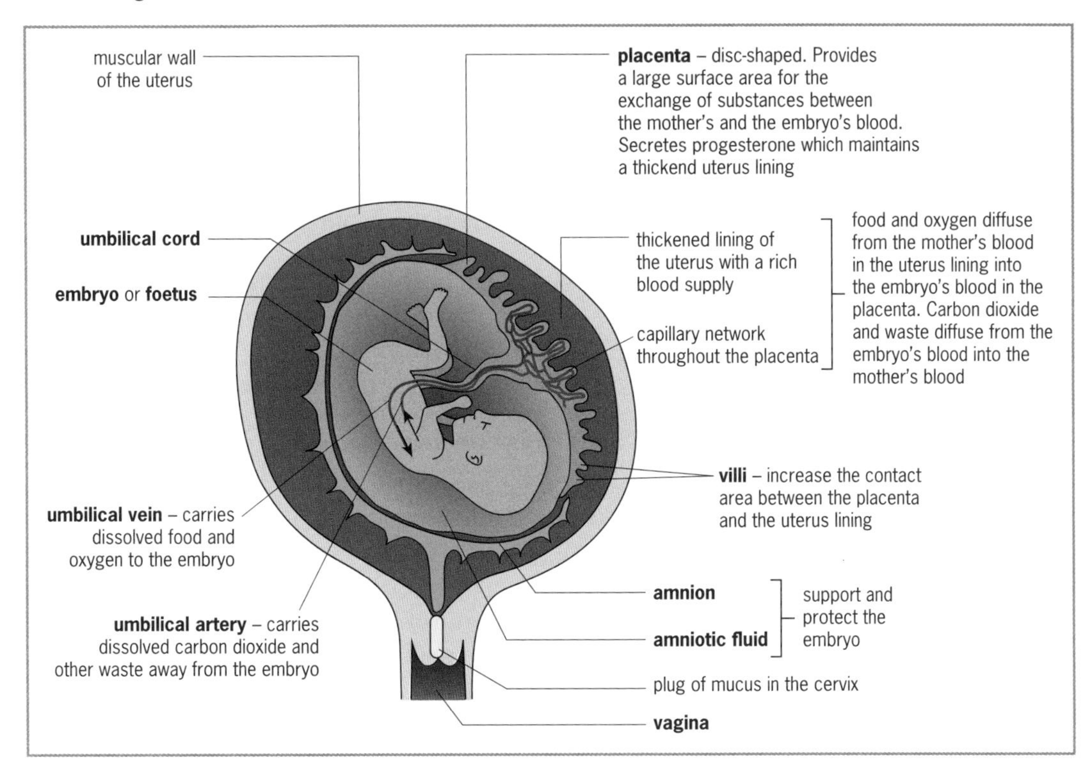

Figure 14.5 *The developing human embryo/foetus in the uterus*

Table 14.3 *The development of a human embryo/foetus*

Time after fertilisation	Characteristics
7 to 10 days	A hollow ball of cells that is implanted in the uterus lining.
4 weeks	The brain, eyes and ears are developing along with the nervous, digestive and respiratory systems. Limb buds are forming and the heart is beginning to beat.
8 weeks	The embryo has a distinctly human appearance. All the vital organs have been formed and limbs with fingers and toes are developed.
10 weeks	The embryo is now known as a **foetus**. External genitals are beginning to appear, fingernails and toenails form and the kidneys start to function.
11 to 38 weeks	The foetus continues to grow and the organs continue to develop and mature.
38 weeks	Birth occurs.

N.B. The **gestation period (pregnancy)** is considered to last for 40 weeks or 280 days since it is calculated from the first day of the last menstrual cycle and not from the time of fertilisation.

Birth

The foetus turns so it lies head down. Secretion of progesterone by the placenta is reduced and this stimulates the **pituitary gland** to secrete the hormone **oxytocin**. Oxytocin stimulates muscles in the uterus wall to start contracting, i.e. **labour** begins. The amnion bursts and the contractions cause the cervix to dilate. When fully dilated the baby is pushed, head first, through the cervix and vagina. The umbilical cord is cut and the placenta is expelled as the **afterbirth** by further contractions of the uterus wall.

The effect of pregnancy on the menstrual cycle

If fertilisation takes place, the **corpus luteum** remains in the ovary and it secretes increasing amounts of **progesterone**. This causes the uterus lining to increase in thickness and it prevents menstruation. As the **placenta** develops, it takes over secreting progesterone which keeps the uterus lining thick and inhibits ovulation and menstruation throughout pregnancy.

Methods of birth control (contraception)

A variety of methods are available to prevent pregnancy from occurring. They are designed to **prevent fertilisation** or to **prevent implantation**. Two methods, **abstinence** and the **condom**, also protect against the spread of sexually transmitted infections (STIs), e.g. AIDS. When choosing a method, its reliability, availability, side effects and whether both partners are comfortable using it, must be considered.

Table 14.4 *Methods of birth control*

Method	How the method works	Advantages	Disadvantages
Abstinence	Refraining from sexual intercourse.	Completely effective. Protects against sexually transmitted infections.	Relies on self control from both partners.
Withdrawal	Penis is withdrawn before ejaculation.	No artificial device needs to be used or pills taken, therefore, it is acceptable to all religious groups.	Very unreliable since some semen is released before ejaculation. Relies on self control.
Rhythm method	Intercourse is restricted to times when ova should be absent from the oviducts.	No artificial device needs to be used or pills taken, therefore, it is acceptable to all religious groups.	Unreliable since the time of ovulation can vary. Restricts the time when intercourse can occur. Unsuitable for women with an irregular menstrual cycle.
Spermicides	Creams, jellies or foams inserted into the vagina before intercourse. Kill sperm.	Easy to use. Readily available.	Not reliable if used alone, should be used with a condom or diaphragm. May cause irritation or an allergic reaction.

Method	How the method works	Advantages	Disadvantages
Condom	A latex rubber or polyurethane sheath placed over the erect penis or into the female vagina before intercourse. Acts as a **barrier** to prevent sperm entering the female body.	Very reliable if used correctly. Easy to use. Readily available. Protects against sexually transmitted infections.	May reduce sensitivity so interferes with enjoyment. Condoms can tear allowing sperm to enter the vagina. Latex may cause an allergic reaction.
Diaphragm	A dome-shaped latex rubber disc inserted over the cervix before intercourse. Should be used with a spermicide. Acts as a **barrier** to prevent sperm entering the uterus.	Fairly reliable if used correctly. Not felt, therefore, does not interfere with enjoyment. Easy to use once the female is taught.	Must be left in place for 6 hours after intercourse, but no longer than 24 hours. Latex may cause an allergic reaction. May slip out of place if not fitted properly.
Intra-uterine device (IUD or coil)	A T-shaped plastic device, usually containing copper or progesterone, inserted into the uterus by a doctor. Prevents sperm reaching the ova or prevents implantation.	Very reliable. Once fitted, no further action is required except an annual check-up. No need to think further about contraception. Few, if any, side effects.	Must be inserted by a medical practitioner. May cause menstruation to be heavier, longer or more painful.
Contraceptive pill	A **hormone** pill, taken daily, which contains oestrogen and progesterone, or progesterone only. Prevents ovulation. Makes cervical mucus thicker and more difficult for sperm to swim through.	Almost totally reliable if taken daily. Menstruation is lighter, shorter and less painful.	Ceases to be effective if one pill is missed. May cause side effects in some women, especially those who smoke.
Surgical sterilisation	The sperm ducts or oviducts are **surgically** cut and tied off. Prevents sperm leaving the male body or ova passing down the oviducts.	Totally reliable. No need to think further about contraception. No artificial device needs to be used or pills taken.	Usually irreversible.

N.B. One **disadvantage** of all methods except abstinence and condoms is that they do not protect against sexually transmitted infections.

Condoms

Diaphragm

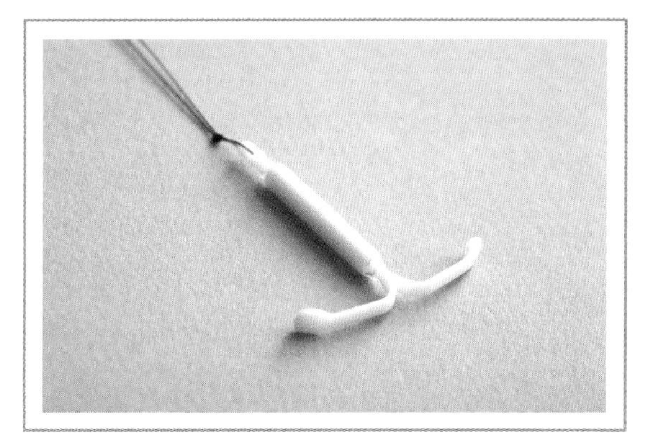

Intra-uterine device (IUD)

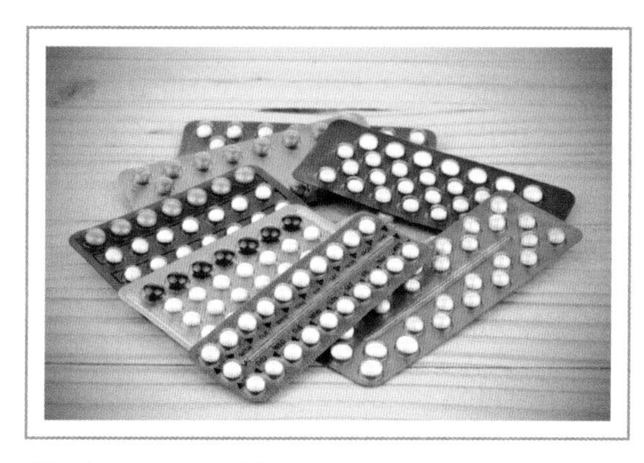

Contraceptive pills

The importance of birth control

Birth control is used to **prevent unintended pregnancies**, therefore, it allows couples to plan when they have children and how many. Access to birth control:

- Decreases health risks to women and maternal deaths caused by unintended pregnancies and unsafe abortions.
- Decreases infant deaths and improves the health and overall care of infants by enabling women to increase the spacing between births.
- Enables family sizes to be limited so each child is provided for emotionally, physically and educationally by having frequent contact with parents.
- Enables women to participate fully in society and advance in the workplace by allowing them to plan for their future and invest in their careers.

From a **global** perspective, the human population is growing rapidly; it is estimated that it will increase from 7.3 billion in 2015 to over 9 billion in 2050. Predictions are that this will result in shortages of food, water, natural resources and land for housing, crops and livestock; will increase pollution, the destruction of the environment, unemployment and the spread of disease; and will decrease living standards. Birth control can **reduce population growth** which should help maintain a healthy, productive environment without shortages, and increase living standards.

Sexually transmitted infections (STIs)

Infections passed on during **sexual intercourse** are called sexually transmitted infections or STIs. These include HIV/AIDS, gonorrhoea, syphilis and genital herpes.

Table 14.5 *Two sexually transmitted infections*

Infection	Causative agent	Methods of transmission	Methods of treatment	Methods of prevention and control
AIDS – acquired immune deficiency syndrome	HIV – human immunodeficiency virus.	• Unprotected sexual intercourse with an infected person. • Using infected hypodermic needles or cutting instruments, e.g. razors. • Transfusions of infected blood products. • Infected mother to baby during pregnancy and breast feeding.	• Antiretroviral drugs to interrupt the duplication of HIV. This keeps the amount of HIV in the body at a low level which delays the onset of opportunistic infections. • Drugs to enhance the immune system. • Drugs to treat opportunistic infections. • No cure exists.	• Treat all cases, especially pregnant women. • Set up education programmes. • Abstain from sexual intercourse or keep to one, uninfected sexual partner. • Use condoms during sexual intercourse. • Trace and treat all sexual contacts of infected persons. • Don't use intravenous drugs or share cutting instruments. • Use sterile needles for all injections. • Test all human products to be given intravenously for HIV. • No vaccine exists.
Gonorrhoea	Bacterium – *Neisseria gonorrhoeae*.	• Unprotected sexual intercourse with an infected person. • Mother to the eyes of her baby during childbirth which leads to blindness if not treated.	• Antibiotics that are specialised to destroy *Neisseria gonorrhoeae*.	• Treat all cases, especially pregnant women. • Set up education programmes. • Abstain from sexual intercourse or keep to one, uninfected sexual partner. • Use condoms during sexual intercourse. • Trace and treat all sexual contacts of infected persons. • No vaccine exists.

Implications of HIV/AIDS

A combination of factors makes the spread of HIV/AIDS very difficult to control:

- There is currently **no vaccine** or **cure** for HIV/AIDS.
- The **time interval** between the virus entering the body and symptoms developing may be several years. During this time the infected person can pass on the virus without knowing it.
- **Highly active antiretroviral therapy (HAART)** where a patient takes a combination of three or more antiretroviral drugs is relatively expensive and must be taken for a patient's lifetime making the cost ongoing.
- It can be difficult to persuade people to change their **sexual behaviour**.

The continual spread of HIV/AIDS has a variety of **consequences** including:

- Shortened life expectancies.
- Job loss resulting in loss of earnings.
- Increased expenditure for medical care and an increased strain on health services.
- Discrimination.
- Neglect by relatives and friends.
- Parentless children.
- Decreased standards of living.

With education, some of these are preventable.

Revision questions

1 **a** Give THREE differences between asexual and sexual reproduction.

b Give ONE advantage and ONE disadvantage of EACH type of reproduction in **a** above.

2 By means of a labelled and annotated diagram, describe the functions of the different parts of the female reproductive system.

3 **a** Describe the events taking place in the ovaries and uterus during one complete menstrual cycle.

b Name FOUR hormones involved in controlling the menstrual cycle.

4 What effect does pregnancy have on the menstrual cycle?

5 Outline the mechanism by which sperm and ova come together to form a zygote in humans.

6 What part does EACH of the following play in the development of a human embryo?

a the amniotic fluid **b** the placenta **c** the umbilical cord.

7 Construct a table that explains how EACH of the following methods of birth control prevents pregnancy, and gives ONE advantage and ONE disadvantage of EACH method: the contraceptive pill, surgical sterilisation, the rhythm method and the condom.

8 Identify THREE ways AIDS is transmitted and THREE ways of controlling its spread.

9 Discuss some of the social implications of the continual spread of AIDS and other sexually transmitted infections.

Sexual reproduction in flowering plants

Flowering plants produce **flowers** for **sexual reproduction**. A flower consists of an expanded stem tip, the **receptacle**, which usually bears four whorls (rings) of modified leaves, **sepals**, **petals**, **stamens**, and one or more **carpels** in the centre.

Most flowers contain both female and male reproductive parts. The female parts are the **carpels**; these produce one or more **ovules** which contain the female gametes. The male parts are the **stamens**; these produce the **pollen grains** which contain the male gametes.

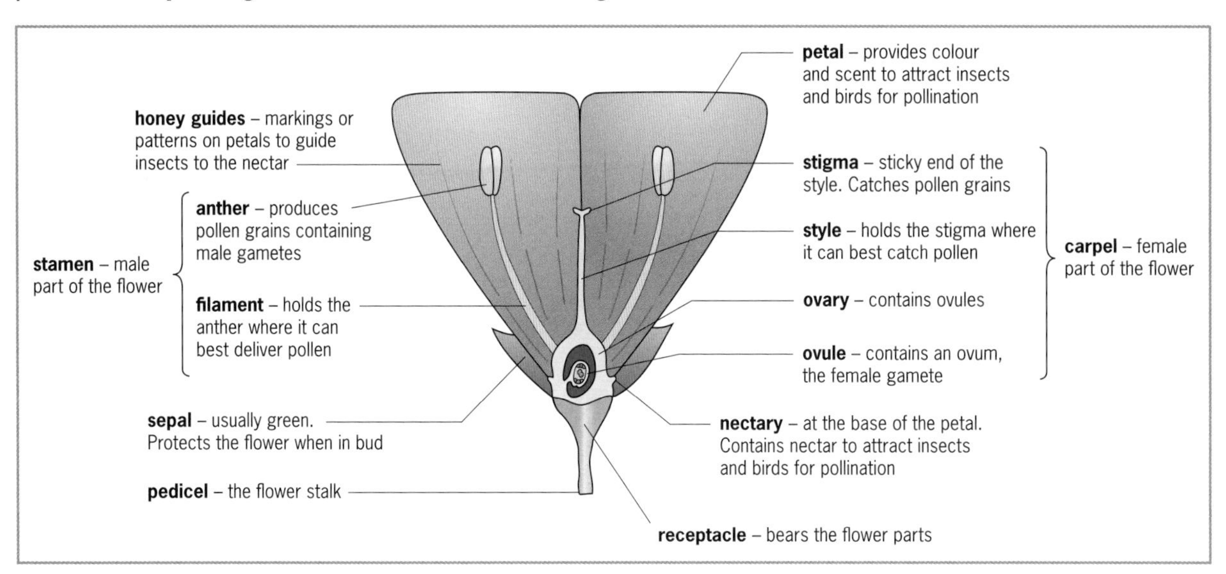

Figure 14.6 *A longitudinal section of a generalised flower showing the function of the parts*

Pollination

***Pollination** is the transfer of pollen grains from the anthers to the stigmas of flowers.*

Pollination leads to **fertilisation** and there are two types:

- **Self pollination** occurs when a pollen grain is transferred from an anther to a stigma of the **same flower** or to a stigma of another flower on the **same plant**.
- **Cross pollination** occurs when a pollen grain is transferred from an anther of a flower on one plant to a stigma of a flower on a **different plant** of the **same species**.

Agents of pollination carry the pollen grains between flowers. They may be the **wind**, **insects** and some **birds**, e.g. humming birds. Flowers are usually **adapted** to be pollinated by wind or by insects.

Table 14.6 *Comparing flowers adapted for wind pollination and insect pollination*

	Wind pollinated	Insect pollinated
Flower	• Usually small and inconspicuous.	• Usually large and conspicuous.
Petals	• Often absent. If present they are small, green or dull coloured and have no scent, nectar or honey guides. There are no pollinating agents to attract.	• Usually relatively large, brightly coloured and scented, and have nectaries and honey guides to attract insects.
Pollen grains	• Small, smooth and light so they are easily carried by the wind. • Large quantities are produced as many are lost.	• Relatively large, sticky or spiky to stick onto the body of insects. • Smaller quantities are produced as fewer are lost.

	Wind pollinated	Insect pollinated
Stamens	• Anthers are loosely attached to long, thin filaments and they hang outside the flower so the pollen can be easily blown off them by the wind. • Anthers are large to produce a lot of pollen grains.	• Anthers are firmly attached to short, stiff filaments, and are usually inside the flower so the insect brushes against them as it goes to get nectar and picks up pollen grains without damaging the anthers.
Stigmas	• Long, branched and feathery and hang outside the flower to provide a large area to catch the pollen grains.	• Flat or lobed and sticky, and are usually situated inside the flower so the insect brushes against them as it goes to get nectar and deposits pollen onto them.
Examples	Guinea grass, maize, sugar cane.	Pride of Barbados, flamboyant, allamanda.

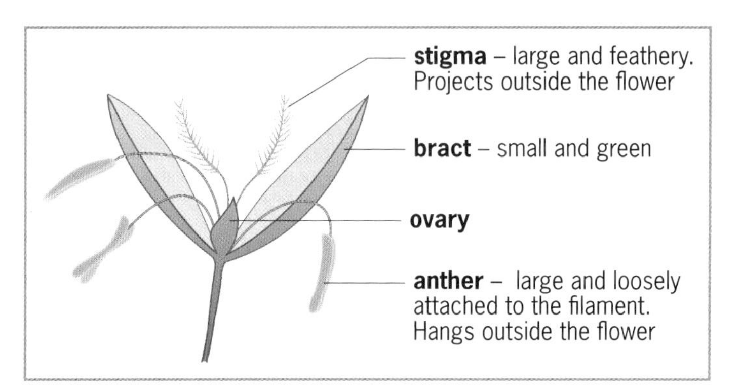

Figure 14.7 *Generalised structure of a wind pollinated flower*

Fertilisation in flowering plants

After pollination has occurred, the male gamete then has to reach the female gamete for **fertilisation** to take place.

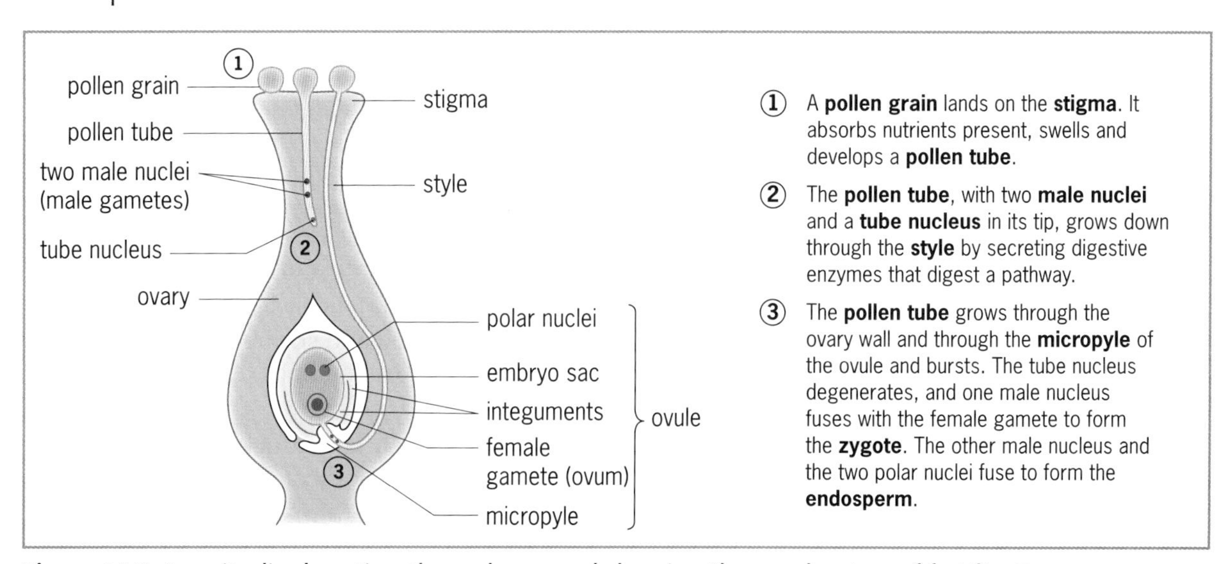

Figure 14.8 *Longitudinal section through a carpel showing the mechanism of fertilisation*

Events following fertilisation in flowering plants

Seed development

After fertilisation, each **ovule** develops into a **seed**:

- The **zygote** divides by **mitosis** forming the **embryo** which develops into three parts:
  - the **plumule** or embryonic shoot
  - the **radicle** or embryonic root
  - **one cotyledon** or seed leaf in monocotyledons or **two cotyledons** in dicotyledons (see Figure 13.2, page 114).
- The **endosperm** remains in **endospermic seeds**, e.g. maize and castor oil, but is absorbed by the cotyledons in **non-endospermic seeds**, e.g. green bean and pigeon pea. In endospermic seeds, the endosperm stores food; in non-endospermic seeds, the cotyledons store food.
- The **integuments** become dry and develop into the **testa**, and the **micropyle** remains in the testa.

Water is withdrawn from the seed and it becomes **dormant**.

Fruit development

After fertilisation, the **ovary wall** develops into the **fruit**. A fruit contains one or more **seeds**; the number depends on the number of ovules in the original ovary that were fertilised. The **shape** and **structure** of many fruits is very similar to the original ovary.

The **stigma**, **style**, **stamens** and **petals** wither and drop off. The **sepals** may drop off or they may remain, e.g. in eggplant.

Fruits

Fruits **protect** the developing seeds and they help to **disperse seeds**. The wall of the fruit is known as the **pericarp** and may be composed of three layers:

- the **exocarp (epicarp)** or outer layer
- the **mesocarp** or middle layer
- the **endocarp** or inner layer.

There are two main **types** of fruits:

- **Succulent (fleshy) fruits.** One or more layers of the pericarp are fleshy and juicy, e.g. mango, guava, tomato and cucumber.
- **Dry fruits.** The pericarp is thin and dry, e.g. the pod of pride of Barbados or pigeon pea and the capsule of castor oil.

A **fruit** has **two** scars, one where it was attached to the parent plant and one where the style was attached. A **seed** only has **one** scar, the hilum, where it was attached to the fruit.

Seed dispersal

Fruits aid in **dispersing** seeds. Spreading seeds away from the parent plant is important to increase the chances of survival.

- Dispersal prevents **overcrowding** thereby preventing competition for light, water, carbon dioxide and minerals.
- Dispersal allows plants to **colonise** new habitats.

Fruits and seeds often have **adaptations** to help dispersal.

- **Dispersal by animals**
  - Many **succulent fruits** contain **stored food** which attracts animals to eat them, e.g. orange, mango, guava, tomato and golden apple.

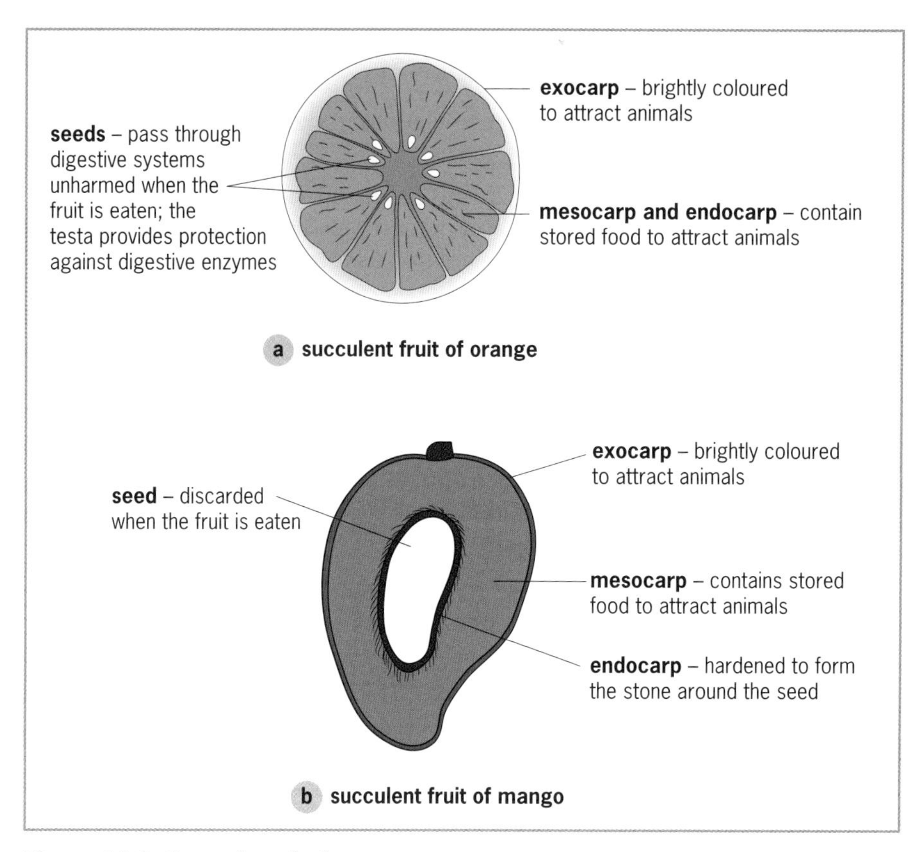

Figure 14.9 *Succulent fruits*

  - Certain small **dry fruits** develop **hooks** that attach the fruits onto the fur of animals, e.g. castor oil, duppy needle, sweethearts and burr grass.

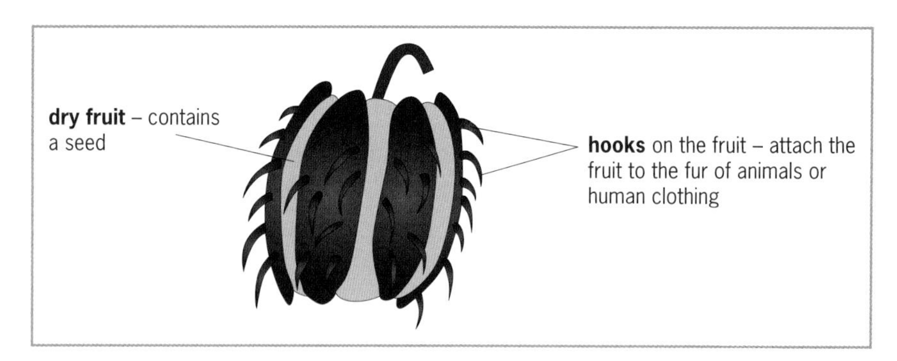

Figure 14.10 *The hooked fruit of castor oil*

- **Dispersal by wind**
  - Some small **dry fruits** develop one or more **wing-like** extensions, e.g. crow and *Combretum*, or the **seeds** contained in certain fruits develop one or more **wings**, e.g. mahogany and *Tecoma*. These provide a large surface area to help the wind carry the fruits or seeds.

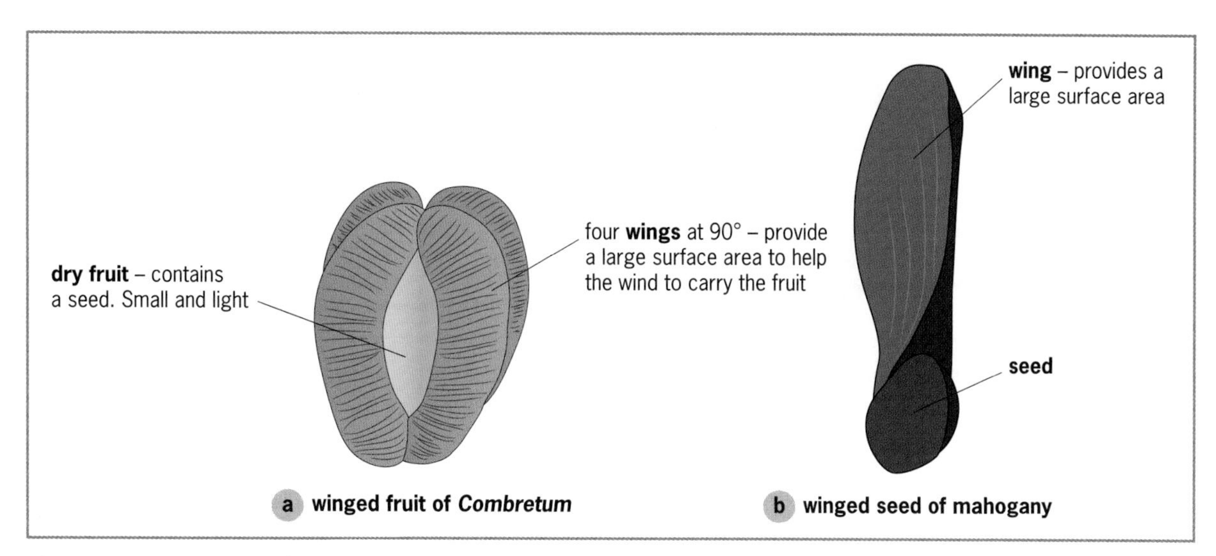

Figure 14.11 *Winged fruits and seeds*

  - Some small **dry fruits** develop **hair-like** extensions that form a 'parachute', e.g. *Tridax*, or the **seeds** contained in certain fruits develop a 'parachute' of **hairs**, e.g. *Stephanotis*, cotton and silk cotton. These provide a large surface area to help the wind carry the fruits or seeds.

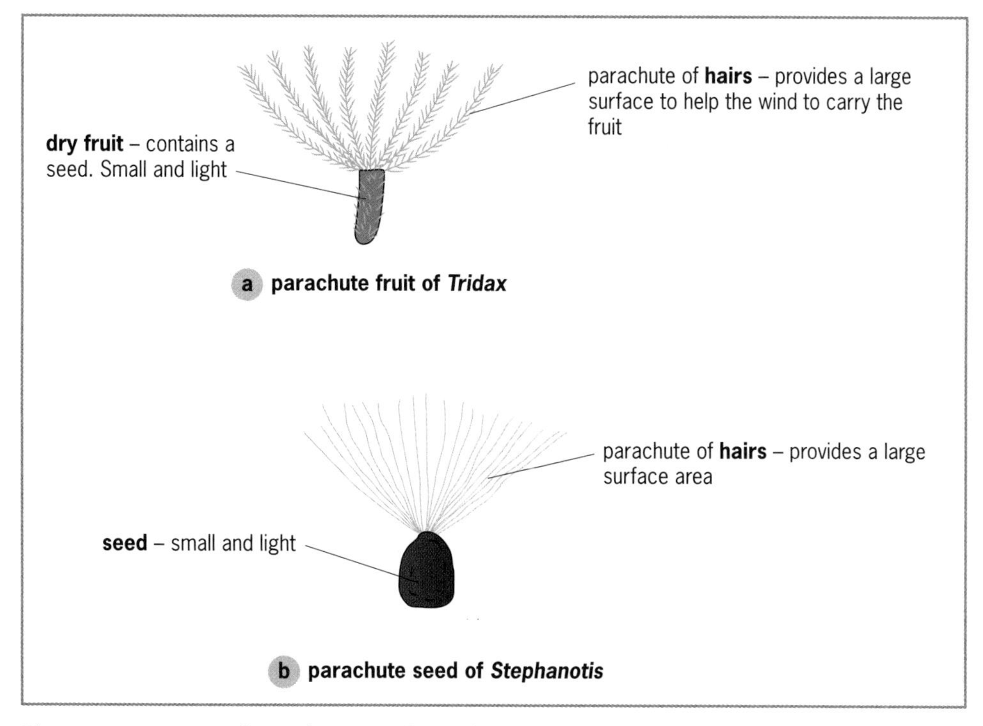

Figure 14.12 *Parachute fruits and seeds*

- **Dispersal by water**

  Some **succulent fruits** develop a waterproof exocarp and become **buoyant** so they can float on water, e.g. coconut and manchineel.

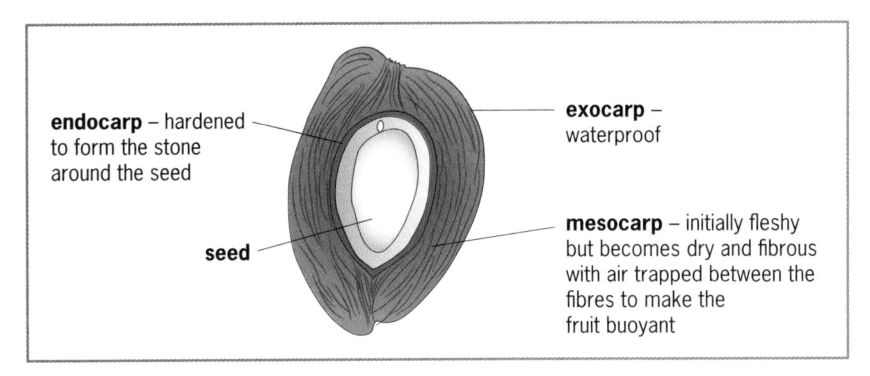

Figure 14.13 *The succulent fruit of coconut*

- **Dispersal by mechanical means**

  Some **dry fruits** split open along lines of weakness and eject their seeds, e.g. pride of Barbados, pigeon pea and crotalaria.

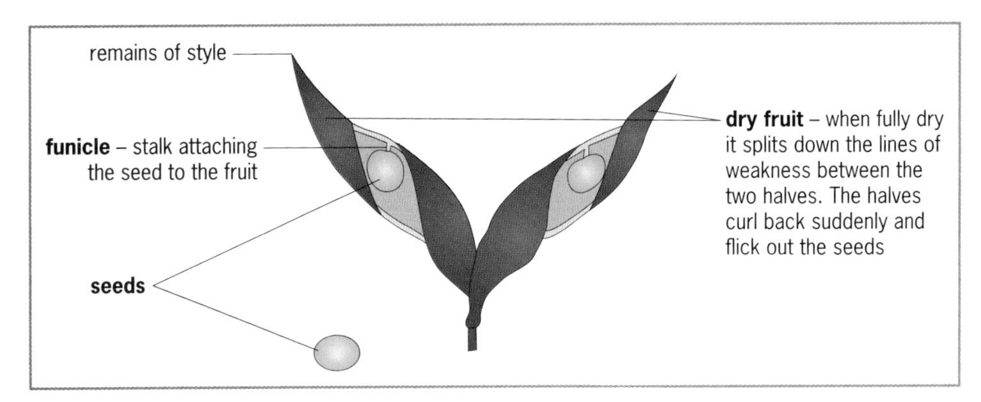

Figure 14.14 *The pod of pride of Barbados*

Revision questions

10 What is the importance of flowers to flowering plants?

11 State the functions of the parts of a flower.

12 What is pollination?

13 Give FOUR features that would enable you to determine that a flower is insect pollinated.

14 Describe the events that occur in the carpel of a flowering plant following pollination that lead to the development of the seed and the fruit.

15 Plants are usually the first organisms to colonise new environments even though they cannot move from place to place by themselves. Explain, giving specific examples, the different ways plants can arrive in new environments.

15 Disease

*A **disease** is a condition that impairs the normal functioning of cells, tissues or organs and it leads to the health of an organism being damaged.*

Diseases can be divided into four main types:

- **pathogenic** diseases
- **deficiency** diseases
- **hereditary** diseases
- **physiological** diseases.

Table 15.1 *The four main types of disease compared*

Type of disease	Cause	Examples
Pathogenic	Microscopic organisms known as **pathogens**. A pathogen is a **parasite** that causes disease in its host, e.g. viruses, bacteria, fungi and protozoans.	AIDS, common cold, influenza, dengue and yellow fever are caused by viruses. Gonorrhoea, syphilis, tuberculosis (TB) and cholera are caused by bacteria. Athlete's foot, thrush and ringworm are caused by fungi. Malaria, amoebic dysentery and sleeping sickness are caused by protozoans.
Deficiency	The shortage or lack of a particular **nutrient** in the diet.	Scurvy is caused by deficiency of vitamin C. Anaemia is caused by a deficiency of iron. Kwashiorkor is caused by a deficiency of protein.
Hereditary	An abnormal **gene** passed on from one generation to the next.	Sickle cell anaemia, cystic fibrosis and Huntingdon's disease.
Physiological	The **malfunctioning** of a body organ or a **change** in the structure of certain body cells over time causing them not to function correctly.	Diabetes, hypertension, cancer and Alzheimer's disease.

Pathogenic diseases are also known as **infectious** or **communicable diseases** because they can be passed on from person to person. Deficiency, hereditary and physiological diseases are also known as **non-communicable diseases** because they cannot be passed on from person to person.

Vectors and the spread of pathogenic diseases

A **vector** is an organism that carries pathogens in or on its body. It transmits the pathogen from one person to another and is not usually harmed by the pathogen, e.g. house flies transmit gastroenteritis, rats transmit leptospirosis, fleas transmit bubonic plague and mosquitoes transmit several diseases.

The life cycle of a mosquito

A mosquito undergoes **complete metamorphosis**, i.e. in growing from young to adult a mosquito passes through **four** distinct **stages**:

- **Egg**: the adult female lays eggs in protected areas that hold water when it rains. The eggs float on the surface of the water.
- **Larva**: the larva hatches from the egg. This is the **feeding** and **growing stage**. Larvae live in the water where they hang from the surface and breathe air through breathing tubes. They feed on micro-organisms and organic matter in the water.
- **Pupa**: the pupa develops from the larva. The pupa is the non-feeding stage in which **larval tissue re-organises** into adult tissue. Pupae live in the water where they hang from the surface and breathe air through two breathing tubes.
- **Adult** or **imago**: the adult emerges from the pupa. The adult is the **flying** and **reproducing stage**. Adults feed on nectar and sugars from plants. After mating, the female requires a **blood meal** to mature her eggs before she lays them. She usually obtains the blood from a human. Adults live in and around human residences where they rest in cool, dark places during the day, and fly and feed in the evenings.

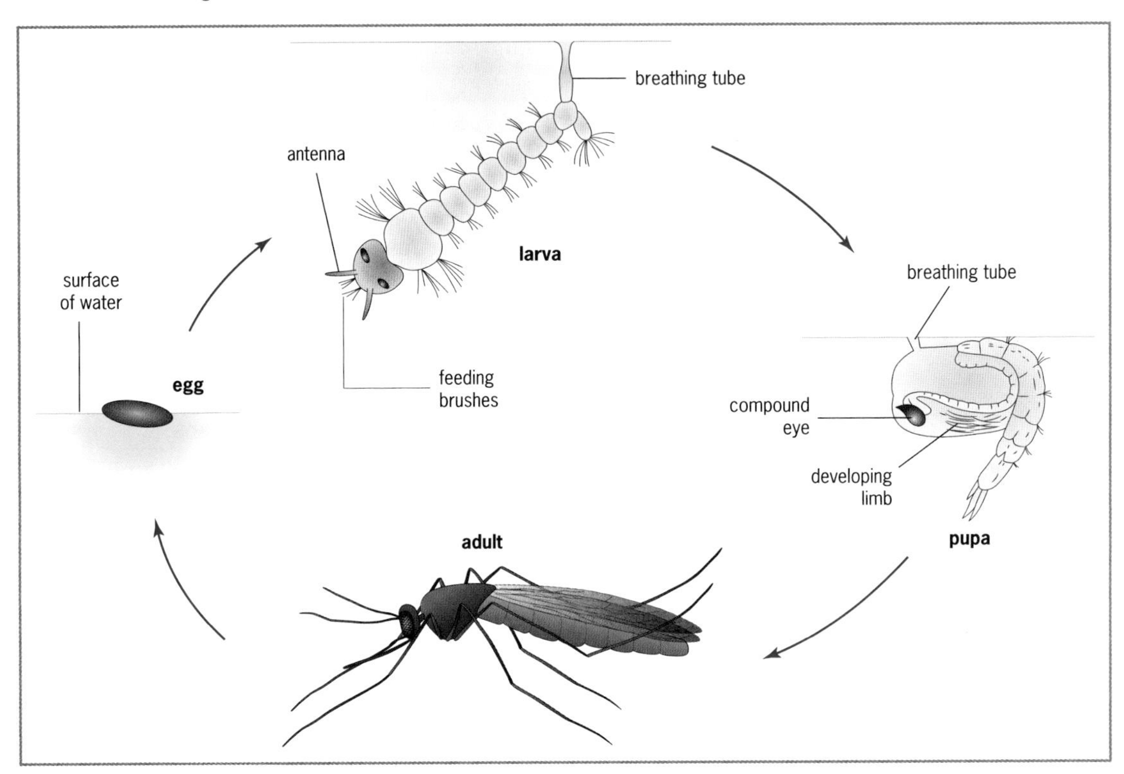

Figure 15.1 *The life cycle of a mosquito*

The role of the mosquito as a vector

Pathogens transmitted by mosquitoes have **two** hosts:

- **Humans** who serve as the **primary host**. The pathogens cause **disease** of the primary host.
- **Mosquitoes** which serve as the **secondary host** or **intermediate host**. The pathogens **reproduce** in the secondary host, but cause it **no harm**.

If the **blood** obtained by a female mosquito to mature her eggs contains pathogens, these pass through the walls of her intestines and move into her **salivary glands** where they remain and **multiply** throughout her lifetime. After laying each batch of eggs she requires a blood meal to mature the next batch. Each time she bites someone to obtain this meal she can transmit the pathogens to that person as she injects saliva into the blood to prevent it clotting before she sucks it up.

Since mosquitoes are unharmed by the pathogens, they serve as **reservoirs** in which the pathogens can multiply and be continually spread to humans. Mosquitoes ensure the continued **survival** and **transmission** of these pathogens.

Diseases transmitted by mosquitoes

The ***Anopheles*** mosquito transmits the protozoan called *Plasmodium* that causes **malaria**, and the ***Aedes*** mosquito transmits the viruses that cause **yellow fever**, **dengue** and **chikungunya**.

- Symptoms of **malaria** include recurrent attacks of suddenly feeling cold and shivering, followed by a high fever and sweating that lasts for several hours. Periodic attacks of high fever and weakness can continue for years. Malaria can be fatal.
- Symptoms of **yellow fever** include a fever, headache, aching muscles, especially back muscles, nausea and vomiting. The toxic phase, occurring in some people, causes yellowing of the skin and eyes, bleeding from the nose, mouth, eyes and internally, and liver and kidney failure. Yellow fever can be fatal.
- Symptoms of **dengue** include a high fever lasting several days, severe headaches, pain behind the eyes, severe joint and muscle pain, nausea, vomiting and a skin rash. **Haemorrhagic dengue** can cause bleeding from the nose, gums, beneath the skin and internally, and can be fatal.
- Symptoms of **chikungunya** are similar to those of dengue with joint pain, which can last for months, particularly affecting wrists, hands, ankles and feet.

The control of mosquitoes

The **control** of vectors is of the utmost importance to control diseases spread by them. To control any vector, its **life cycle** must be understood to work out at which stage or stages control would be the easiest and most effective.

- Mosquito **larvae** and **pupae** can be controlled by:
  - **Draining** all areas of standing water.
  - Adding **insecticides** to breeding areas to kill the larvae and pupae.
  - Introducing **fish** such as *Tilapia* into breeding areas to feed on the larvae and pupae, i.e. biological control.
  - Spraying **oil**, **kerosene** or non-toxic **lecithins** onto still-water breeding areas to preventing the larvae and pupae from breathing.
- **Adult** mosquitoes can be controlled by:
  - Removing **dense vegetation** to reduce protection for adults during daylight hours.
  - Spraying with **insecticides** to kill the adults.

Treatment and control of disease

The aim of **treating** a disease is to **relieve the symptoms** experienced by persons suffering from the disease and **cure** the disease if possible. The aim of **controlling** a disease is to **prevent further development** and **spread** of the disease so that the incidence of the disease in the population is gradually reduced. Treating a disease is always one method to control it. The ultimate goal of treating and controlling any disease is to totally **eradicate** it from the human population.

Table 15.2 *Methods used to treat the four types of disease*

Type of disease	Treatment methods
Pathogenic diseases	Drugs to relieve symptoms, e.g. aspirin to reduce fever and pain killers to reduce pain. Drugs, creams and ointments to kill the pathogens: • antibiotics to kill bacteria • antiviral agents to kill viruses • antifungal agents to kill fungi. Injections of ready-made antibodies to destroy pathogens, e.g. antibodies against tetanus are used for a 'tetanus prone' wound.

Type of disease	Treatment methods
Deficiency diseases	A diet containing foods rich in the missing nutrient or foods fortified with the missing nutrient. Dietary supplements containing the missing nutrient.
Hereditary diseases	Drugs to relieve symptoms as they develop. No cure currently exists for any hereditary disease.
Physiological diseases	Drugs to relieve symptoms as they develop. Other treatments specific to the disease (see below, for diabetes and hypertension).

Table 15.3 *Methods used to control the four types of disease*

Type of disease	Control methods
Pathogenic diseases	Quarantine (isolate) and treat contagious individuals. Set up immunisation programmes and ensure all individuals in populations are vaccinated. Eradicate vectors of disease. Improve sanitation and sewage treatment. Ensure drinking water is properly treated. Set up public health education programmes and improve public health. Practice good personal hygiene and food preparation techniques. Use condoms during sexual intercourse to control STIs.
Deficiency diseases	Improve nutrition within populations. Set up public health education programmes that focus on nutrition.
Hereditary diseases	Avoid situations that worsen the symptoms. Genetic counselling to predict the likelihood that offspring will develop the disease.
Physiological diseases	Adopt a lifestyle that reduces exposure to risk factors. Eat a healthy, balanced diet with plenty of roughage, fresh fruit and vegetables. Reduce obesity. Take regular, moderate exercise to maintain fitness. Attend regular check-ups with a doctor.

Physiological diseases

- **Diabetes** is a disease where blood glucose levels cannot be properly regulated:
  - Type 1 diabetes is caused by the pancreas not producing any insulin. It is treated by regular injections of **insulin**. Blood glucose levels of people with type 1 diabetes can suddenly drop too low, in which case an injection of **glucagon** can be given (see page 62).
  - Type 2 diabetes is caused by the pancreas not producing enough insulin or by the body cells not responding to the insulin. It is usually treated by taking tablets to lower blood glucose levels.
- **Hypertension** or **high blood pressure** is a condition in which the pressure of the blood in the arteries is higher than normal. Factors that contribute to its development include being overweight or obese, smoking, too much salt or too much fat in the diet, consumption of too much alcohol, lack of physical exercise and stress. It is treated by taking drugs to lower blood pressure.

The role of diet in controlling physiological diseases

Diabetes and **hypertension** can both be controlled by eating a healthy, balanced diet (see page 55).

The role of exercise in controlling physiological diseases

- **Diabetes** can be controlled by regular, moderate aerobic exercise such as swimming, walking and aerobics. Exercise increases muscular activity and reduces blood glucose levels by increasing respiration in muscle cells. It also reduces obesity, improves circulation and maintains fitness.
- **Hypertension** can be controlled by plenty of regular, moderate aerobic exercise to reduce obesity and stress, improve circulation and maintain fitness; at least 30 minutes daily is recommended.

Social, environmental and economic implications of disease

Disease within **human populations** can cause loss of earnings as persons with the disease are unable to work. Businesses then become less productive due to a reduction in hours of labour, and this leads to a reduced economy. Demands on health services increase as more people have to seek treatment. Ultimately, human resources are lost and standards of living are reduced.

Disease within **livestock** and **agricultural crops** results in decreased or lost food production, loss of income for the farmer, and a reduced economy, especially if the produce was for export. It also leads to decreased food availability, increased food prices on the local market and reduced standards of living of those whose livelihoods depend on agriculture.

Revision questions

1. Distinguish among a pathogenic disease, a deficiency disease, a hereditary disease and a physiological disease, and give a named example of each.
2. What is a vector?
3. Describe the life cycle of a mosquito, indicating the habitat and mode of life of each stage.
4. Outline the role played by the mosquito in transmitting pathogenic diseases. Include the names of THREE diseases transmitted by mosquitoes.
5. Suggest the most appropriate methods to control the different stages of the life cycle of a mosquito.
6. What differences exist between the methods used to treat diseases caused by pathogens, deficiency diseases and physiological diseases?
7. Discuss the role of BOTH diet and exercise in controlling:

   a diabetes **b** hypertension.
8. Outline some of the consequences of an outbreak of a disease within the human population of your country.

Exam-style questions – Chapters 5 to 15

Structured questions

1 **a)** Figure 1 shows a generalised plant cell.

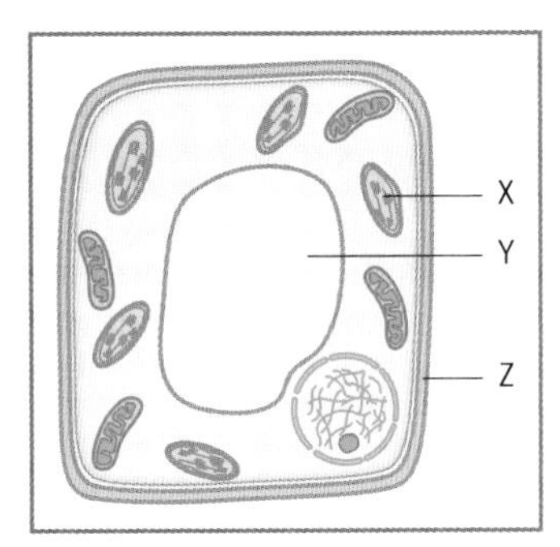

Figure 1 *A generalised plant cell*

- **i)** Identify the structures labelled X and Y. **(2 marks)**
- **ii)** State ONE function of X and TWO functions of Y. **(3 marks)**
- **iii)** In what way does the property of structure Z differ from the cell membrane? **(2 marks)**
- **iv)** Describe the appearance of the cell shown above if it was placed in a concentrated sodium chloride solution and left for 30 minutes. **(2 marks)**
- **v)** Account for your observations in **iv**) above. **(3 marks)**

b) The nucleus is lost in a few types of cells as they mature.

- **i)** What function of the cell is immediately lost when this occurs? **(1 mark)**
- **ii)** Suggest, with an example, ONE other consequence that might result from the lack of a nucleus. **(2 marks)**

Total 15 marks

2 **a)** Figure 2 shows a transverse section through a dicotyledon leaf.

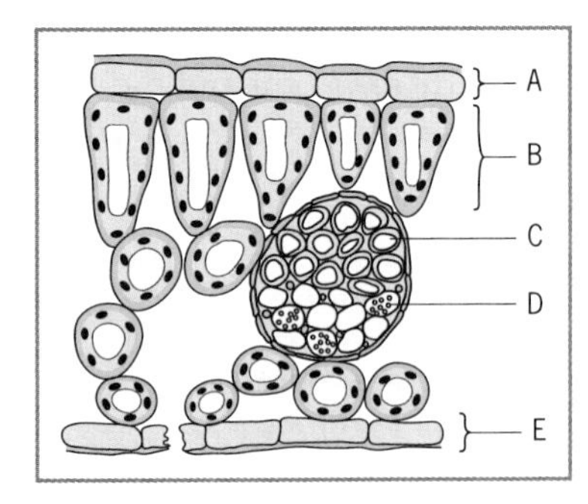

Figure 2 *A transverse section of a dicotyledon leaf*

- **i)** Name the structures labelled B, C and D. **(3 marks)**
- **ii)** Name the main process that takes place in a leaf. **(1 mark)**
- **iii)** Identify TWO differences, shown in Figure 2, between the structure of A and the structure of E, and explain why EACH difference is important. **(4 marks)**

b) Root tubers of the yam plant serve as underground storage organs.

i) Name the main food stored by these root tubers. **(1 mark)**

ii) Describe the test you would carry out to prove the presence of the food named in **i)** above. **(2 marks)**

iii) Explain how the food named in **i)** above reached these underground root tubers. **(4 marks)**

Total 15 marks

3 **a)** Figure 3 shows a longitudinal section through the human heart.

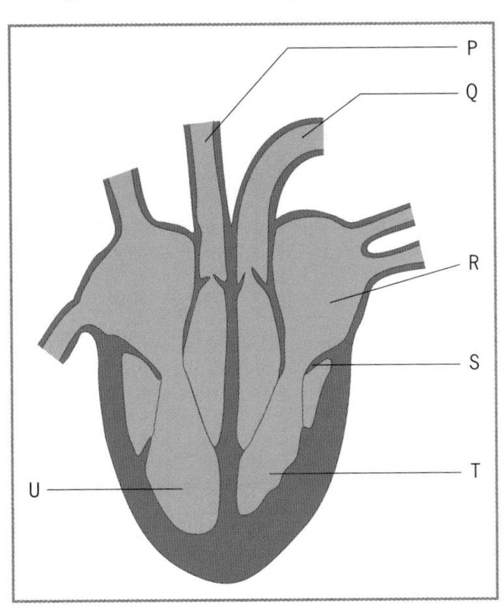

Figure 3 *A longitudinal section through the human heart*

i) Name the structures labelled P to S. **(4 marks)**

ii) State TWO ways in which the composition of the blood in chamber T differs from the composition of the blood in chamber U. **(2 marks)**

iii) Explain why the wall of chamber T is thicker than the wall of chamber U. **(3 marks)**

b) **i)** Suggest TWO reasons why blood plasma is a good transport medium. **(2 marks)**

ii) Suggest TWO reasons why it is dangerous for a person to lose too much blood. **(2 marks)**

c) Give TWO major differences between transport in humans and transport in green plants. **(2 marks)**

Total 15 marks

4 **a)** Figure 4 is a diagram of a knee joint.

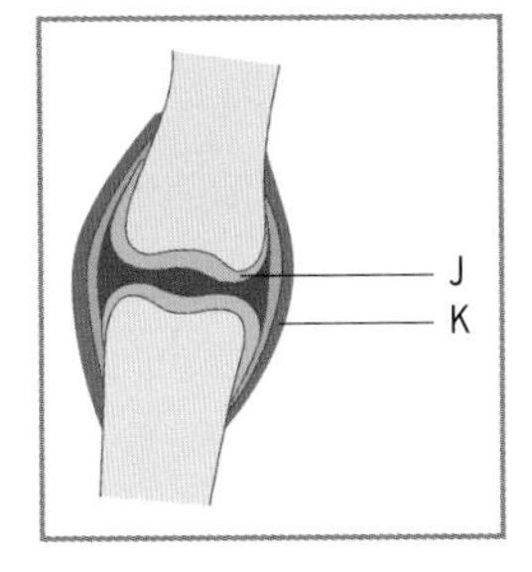

Figure 4 *A knee joint*

i) Name the type of joint shown in Figure 4. **(1 mark)**

ii) Identify the structures labelled J and K and give ONE function of EACH structure. **(4 marks)**

iii) Why does the knee joint require two muscles for movement to occur? **(2 marks)**

iv) State TWO ways in which the joint shown in Figure 4 and the joint found at the hip differ in terms of function. **(2 marks)**

v) As some people age, they find that they need to replace one or both knee joints. Suggest TWO reasons why this might be necessary. **(2 marks)**

b) The veins of a leaf are often referred to as the skeleton of the leaf.

i) Name the TWO types of tissue found in these veins. **(2 marks)**

ii) Explain how these veins act as the leaf's skeleton. **(2 marks)**

Total 15 marks

5 **a)** Figure 5 shows the male reproductive system.

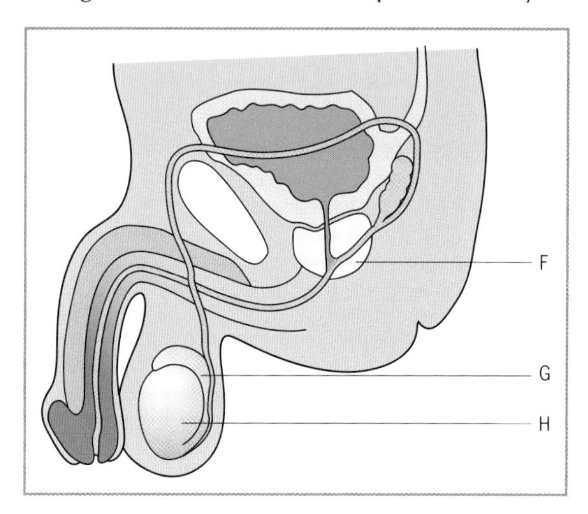

Figure 5 *The male reproductive system*

i) Name the parts labelled F, G and H and give ONE function of EACH part. **(6 marks)**

ii) Place arrows on Figure 5 to show the pathway that sperm takes as it travels out of the male body. **(2 marks)**

iii) How does the urethra in the male body differ in function from the urethra in the female body? **(1 mark)**

b) Figure 6 below shows the structure of a flower.

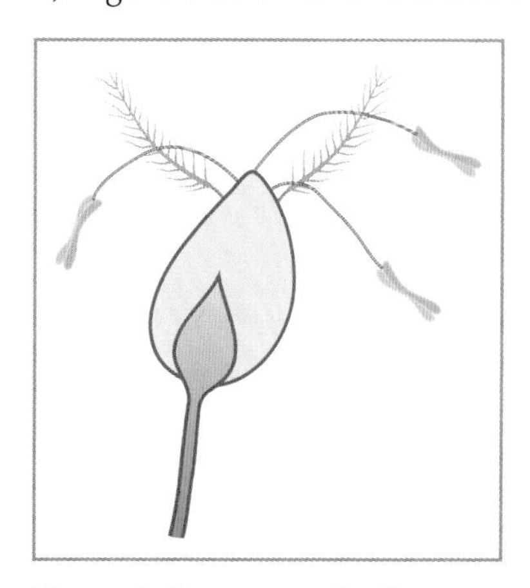

Figure 6 *Structure of a flower*

i) What is likely to be the pollinating agent of the flower shown in Figure 6? **(1 mark)**

ii) Based on the information shown in Figure 6, give TWO reasons to support your answer to **i)** above. **(2 marks)**

iii) Suggest ONE other characteristic that the flower in Figure 6 might have which is useful in pollination. **(1 mark)**

c) Give the main difference between the way in which gametes are brought together in flowering plants and the way they are brought together in humans. **(2 marks)**

Total 15 marks

Extended response questions

6 a) Breast milk is described as a complete food for babies.

i) What should breast milk contain if the above statement is true? **(2 marks)**

ii) Describe what happens to this milk as it passes through a baby's digestive system until it reaches her ileum. **(6 marks)**

iii) There is a tendency for babies fed on powdered milk to gain weight more rapidly than breast fed babies. Suggest TWO reasons for this. **(2 marks)**

b) Babies must develop teeth before they can properly digest solid foods.

i) Explain why teeth are necessary to help digest solid foods. **(3 marks)**

ii) Why is it important that the enamel of teeth is the hardest substance found in the human body? **(2 marks)**

Total 15 marks

7 a) i) Explain why both breathing and gaseous exchange are essential to organisms that respire aerobically. **(2 marks)**

ii) By means of a fully labelled and annotated diagram <u>only</u>, describe the structure of a human lung. **(6 marks)**

b) i) Identify the gaseous exchange surface in a human and explain how it is adapted to efficiently exchange gases. **(5 marks)**

ii) Suggest TWO ways in which cigarette smoking can reduce the efficiency of gaseous exchange in the lungs. **(2 marks)**

Total 15 marks

8 a) i) Distinguish between excretion and egestion. **(2 marks)**

ii) Draw a simple diagram to show a kidney tubule and its blood supply. Annotate your diagram to show how urine is produced. **(7 marks)**

b) You spend most of the day playing tennis but do not drink anything, even though the weather is hot. What effect would this behaviour have on the quantity and composition of your urine? Explain your answer. **(6 marks)**

Total 15 marks

9 a) i) Distinguish between a receptor and an effector. **(2 marks)**

ii) A mother sees her 3-year-old son fall down and hurt himself. She immediately runs to his assistance. Outline the events occurring in her nervous system to bring about her response. **(5 marks)**

b) Discuss the effects of alcohol abuse in humans. Your answer should include:

– a discussion of the effects of short-term abuse on the body's ability to respond to stimuli

– reference to TWO effects of long-term abuse on organs of the body

– a discussion of FOUR social and/or economic effects of alcohol abuse on society. **(8 marks)**

Total 15 marks

10 **a)** Certain pathogenic diseases are spread by vectors.

i) What is a vector? **(2 marks)**

ii) Using diagrams, identify the main stages in the life cycle of a mosquito and indicate where EACH stage would be found. **(6 marks)**

iii) How can the knowledge of the life cycle of a mosquito help prevent the spread of a named pathogenic disease? **(2 marks)**

b) Your father is planning to travel to a country that has recently had an outbreak of cholera, a disease caused by bacteria. Explain how a vaccine can help prevent your father from becoming infected with the cholera. **(5 marks)**

Total 15 marks

16 Inheritance and variation

No two living organisms are identical, they all show **variation**. Much of this variation is passed on from one generation to the next via **genes**.

An introduction to chromosomes and genes

Chromosomes

Chromosomes are present in the nuclei of all living cells. Each chromosome is composed of a single **deoxyribonucleic acid (DNA) molecule** wrapped around **proteins** called **histones**. DNA molecules contain **genetic information** in the form of **genes**. In any cell that is not dividing, chromosomes exist as long, thin strands known as **chromatin threads** which are spread throughout the nucleus. Chromosomes become visible when a cell begins to divide due to them becoming shorter and thicker.

Chromosomes are passed on from one generation to the next in **gametes** and each species has a distinctive **number** of chromosomes per body cell, for example, every human cell has 46 chromosomes. The number of chromosomes in each cell is known as the **diploid number** or **2*n* number**. Chromosomes exist in pairs known as **homologous pairs**. Every human cell has 23 pairs, one member of each pair being of **maternal origin** and the other of **paternal origin**. With the exception of the pair of sex chromosomes, members of each pair look alike.

Genes

Genes are specific sections of chromosomal DNA molecules and are the basic units of **hereditary**. Each human body cell has over 30,000 genes and each gene controls a particular characteristic. Genes work by controlling the production of **protein** in cells, mainly the production of **enzymes**. Each gene controls the production of a specific protein.

All the cells of one organism contain an **identical combination** of genes. It is this combination that makes each organism **unique** since no two organisms, except identical twins or organisms produced asexually from one parent, have the same combination of genes. Within any cell some genes are active while others are inactive, e.g. in a nerve cell, genes controlling the activity of the nerve cell are active and genes that would control the activity of a muscle cell are inactive.

Cell division

When a cell divides, **chromosomes** with their **genes** are passed on to the cells produced, known as **daughter cells**. There are two types of cell division, **mitosis** and **meiosis**.

Mitosis

Mitosis occurs in **all body cells** except in the formation of gametes. During mitosis, **two genetically identical** cells are formed. Each cell contains the **same** number of chromosomes as the parent cell, i.e. the **diploid number**.

Mitosis is important because:

- It ensures that each daughter cell contains the **diploid** number of chromosomes. This maintains the **species number** of chromosomes in all members of a species.
- It ensures that each daughter cell has an **identical** combination of genes.
- It is the method by which all cells of a multicellular organism are formed, hence it is essential for **growth** and to **repair** damaged tissues.
- It is the method by which organisms reproduce **asexually** forming offspring that are **identical** to each other and to the parent.

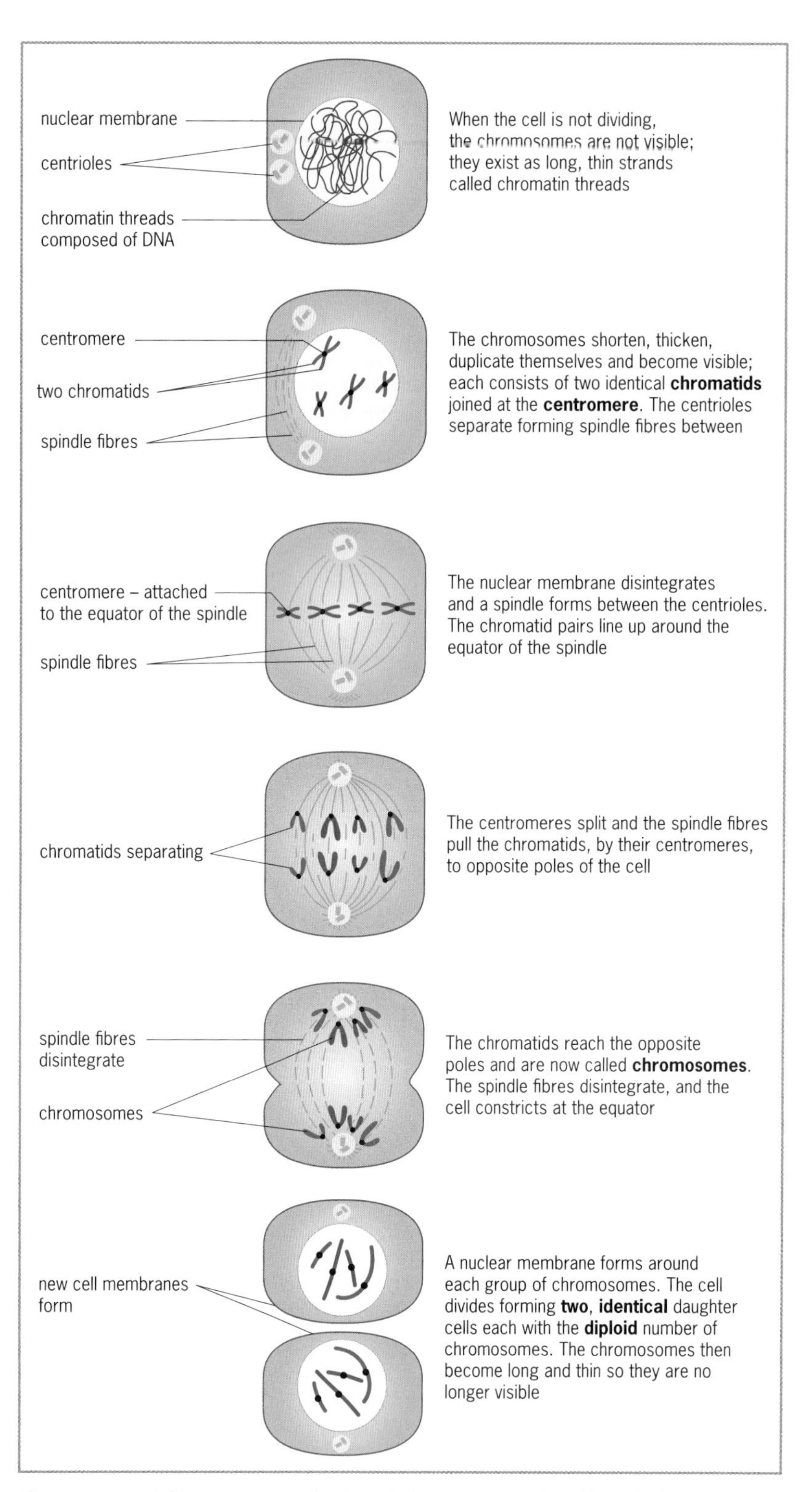

Figure 16.1 *The process of mitosis in an animal cell with four chromosomes; two of paternal origin (blue) and two of maternal origin (red)*

Mitosis and asexual reproduction in plants

Some plants can reproduce **asexually** by **mitosis** occurring in certain structures of the parent plant, a process known as **vegetative propagation**. Since mitosis produces genetically identical cells, all offspring produced asexually from one parent are **genetically identical** and are collectively called a **clone. Cloning** is the process of making genetically identical organisms through non-sexual means.

Examples of natural vegetative propagation

- New plants can grow from **vegetative organs** at the beginning of the growing season, e.g. from rhizomes, stem tubers, corms and bulbs (see page 89).
- New plants can grow from **outgrowths** of the parent plant, e.g. from runners, leaf buds and suckers.

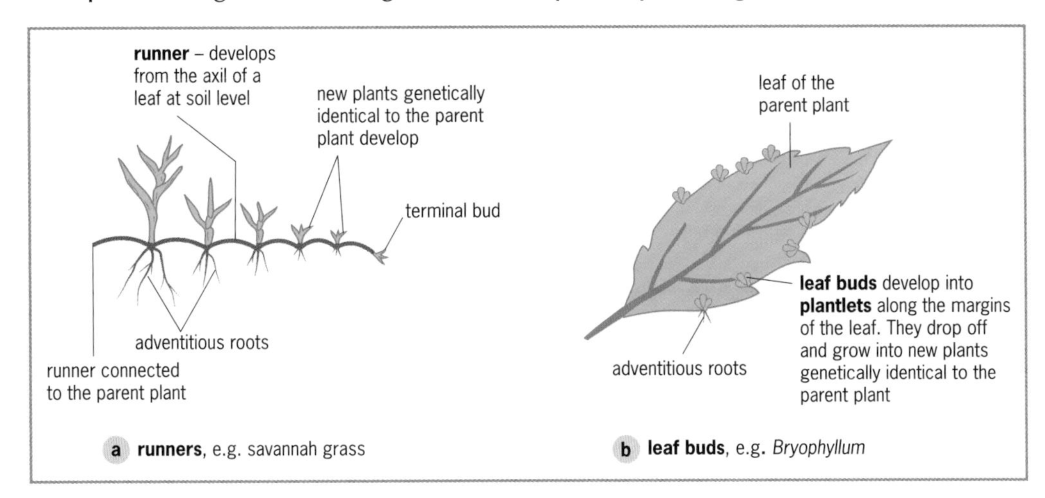

Figure 16.2 *Examples of vegetative propagation in plants*

Examples of artificial vegetative propagation

- By taking **cuttings**, farmers and gardeners artificially propagate plants. Cuttings are parts of plants that will develop roots and shoots to become new plants if given suitable conditions, e.g. **stem cuttings**.
  - When a piece of a **sugar cane** stem with two or three buds is placed horizontally on the soil, new plants grow from each bud.
  - When a stem of **hibiscus** plant with a few leaves at the top is planted, roots grow from the cut end forming a new plant.
- **Tissue culture** is used to artificially propagate plants, e.g. to propagate orchids, potatoes and tomatoes. Small pieces of tissue called **explants** are taken from a parent plant and grown in a nutrient-rich culture, under sterile conditions, to form cell masses known as **calluses**. Each callus is then stimulated with appropriate plant hormones to grow into a new plant.

If cuttings or explants are taken from plants with **desirable characteristics**, e.g. a high yield, high quality, resistance to disease or fast growth rate, then all plants produced will have the same desirable characteristics.

Cloning in animals

To **clone** an animal, a nucleus is removed from an ovum of a female donor. A cell, still containing its nucleus, is taken from the animal to be cloned and is fused with the ovum. This newly created ovum is placed into a surrogate mother where it is stimulated to develop into an embryo. The surrogate then gives birth to a new individual that is **genetically identical** to the animal from which the original cell came, e.g. Dolly the sheep. A very low percentage of cloned embryos survive to birth, and animals born alive often have health problems or other abnormalities, and reduced life spans.

Meiosis (reduction division)

Meiosis occurs only in the **reproductive organs** during the production of **gametes**. During meiosis, **four genetically non-identical** cells are formed. Each cell contains **half** the number of chromosomes as the parent cell, known as the **haploid number** or ***n*** **number**.

Meiosis is important because it ensures that:

- Each daughter cell has the **haploid** number of chromosomes. The **diploid** number can then be restored at **fertilisation**.

- Each daughter cell has a **different** combination of genes. This leads to **variation** among offspring which enables species to constantly change and adapt to changing environmental conditions (see page 156).

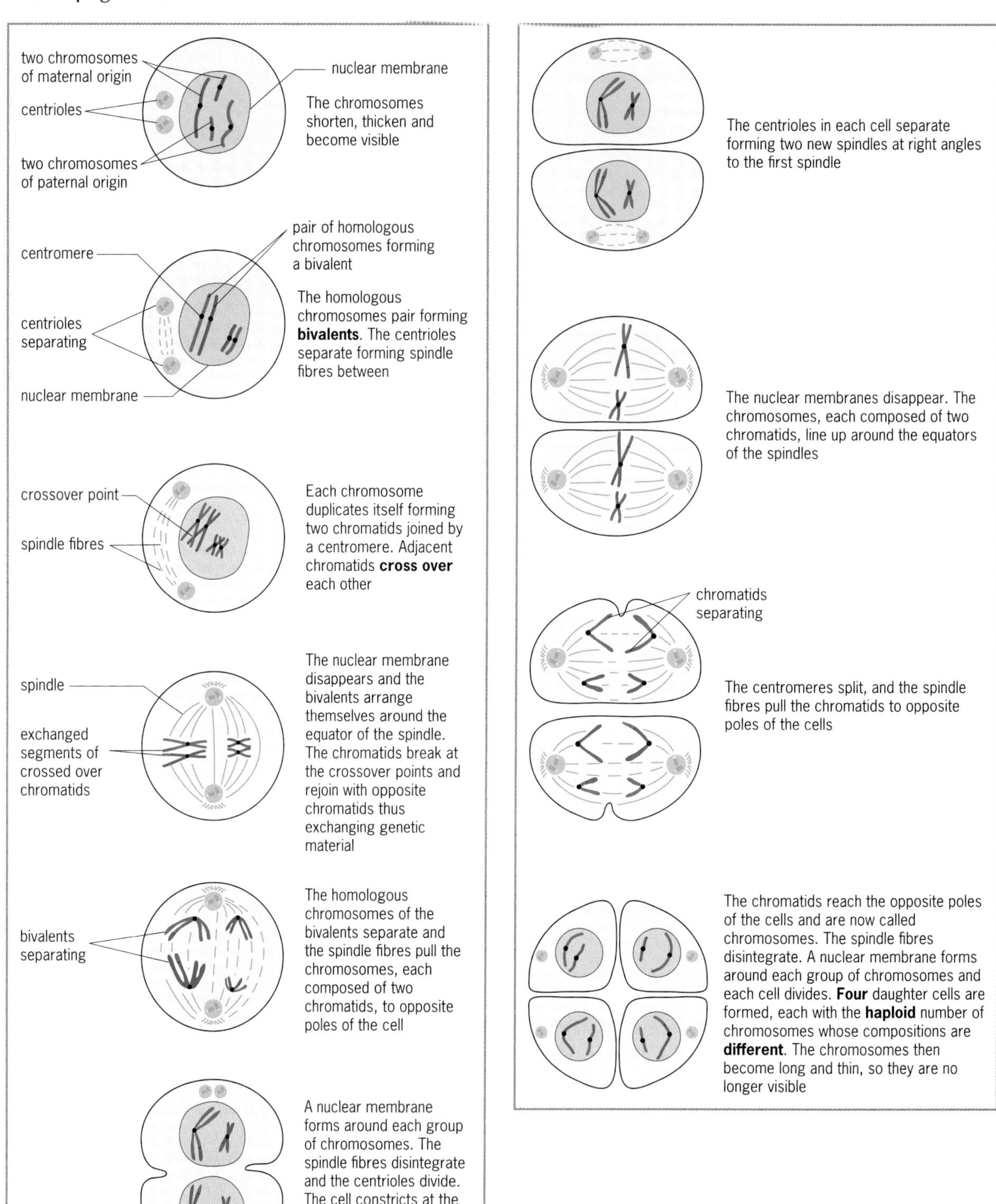

Figure 16.3 *The process of meiosis in an animal cell with four chromosomes*

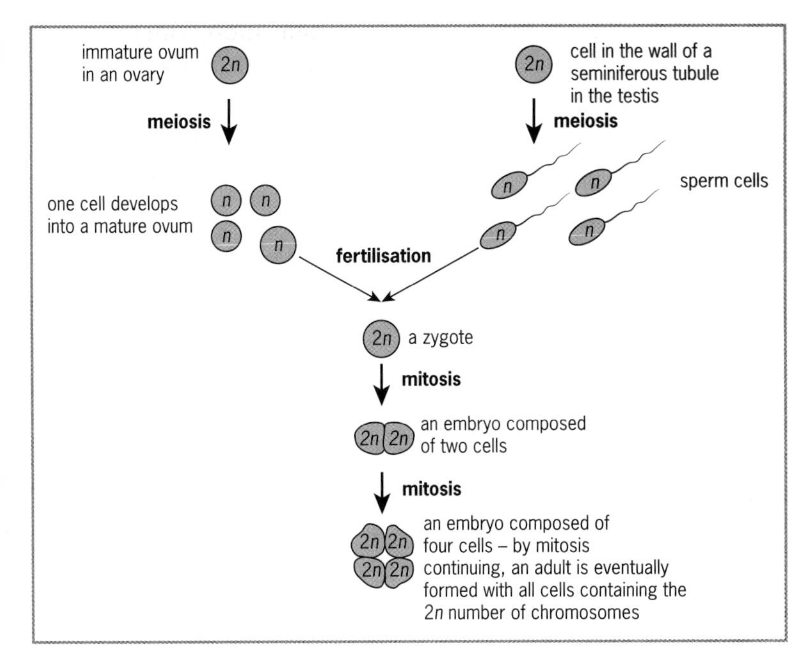

Figure 16.4 *The relationship between mitosis and meiosis*

Revision questions

1. Outline the relationship between chromosomes and genes.
2. Outline the process of mitosis in an animal cell.
3. Give THREE reasons why mitosis is important to living organisms.
4. Describe TWO different natural ways that plants can reproduce asexually.
5. Where does meiosis occur in living organisms?
6. In what ways does meiosis differ from mitosis?
7. Give TWO reasons why meiosis is important to living organisms.

Inheritance

Like chromosomes, genes exist in **pairs**. One gene of each pair is of **maternal origin** and one is of **paternal origin**, and the pairs occupy equivalent positions on homologous chromosomes. A gene controlling a particular characteristic can have different forms known as **alleles**. Each gene usually has two different alleles.

The **composition of genes** within the cells of an organism makes up the organism's **genotype**. The **observable characteristics** of an organism make up its **phenotype**.

Example: albinism in humans

People with **albinism** produce very little or no melanin in their skin, eyes and hair. The gene controlling the production of the pigment melanin has **two** different **alleles** which can be represented using letters:

- **N** stimulates melanin production
- **n** fails to stimulate melanin production

The allele stimulating melanin production, **N**, is **dominant**, i.e. if it is present it shows its effect on the phenotype. The allele for albinism, **n**, is **recessive**, i.e. it only has an effect on the phenotype if there is no dominant allele present. Three combinations of these alleles are possible; **NN**, **Nn** and **nn**. If the two alleles are the same, the organism is said to be **homozygous**. If the two alleles are different, the organism is said to be **heterozygous**.

Table 16.1 *Possible combinations of the alleles controlling melanin production*

Genotype (combination of alleles)	**How the alleles appear on homologous chromosomes**	**Phenotype** (appearance)
NN **Homozygous dominant** (pure breeding)	N N	Normal pigmentation of the skin, eyes and hair
Nn **Heterozygous** (carrier)	N n	Normal pigmentation of the skin, eyes and hair
nn **Homozygous recessive** (pure breeding)	n n	Albino – very pale skin that does not tan, white or light blond hair and very pale blue eyes

Gametes produced in meiosis contain only **one chromosome** from each homologous pair. As a result, they contain only **one allele** from each pair. When fertilisation occurs, chromosomes and the alleles they carry **recombine** to form pairs in the zygote.

Results of possible crosses

1 If one parent is **homozygous dominant** and one is **homozygous recessive**:

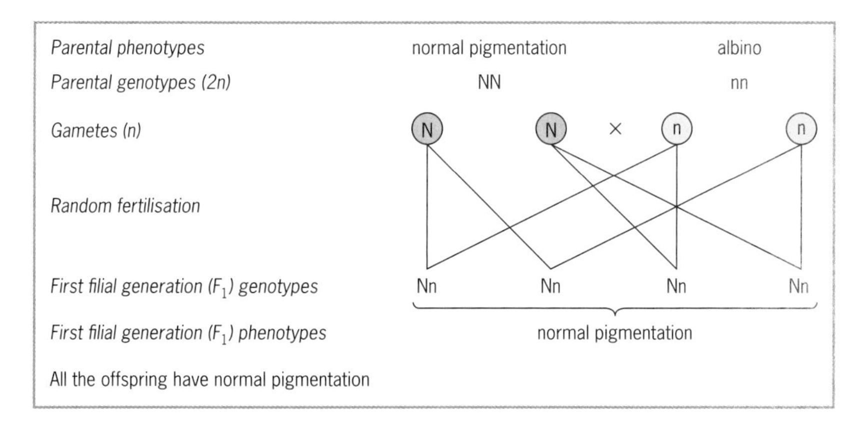

2 If one parent is **heterozygous** and one is **homozygous recessive**, showing the use of a **Punnett square** to predict the outcome of the cross:

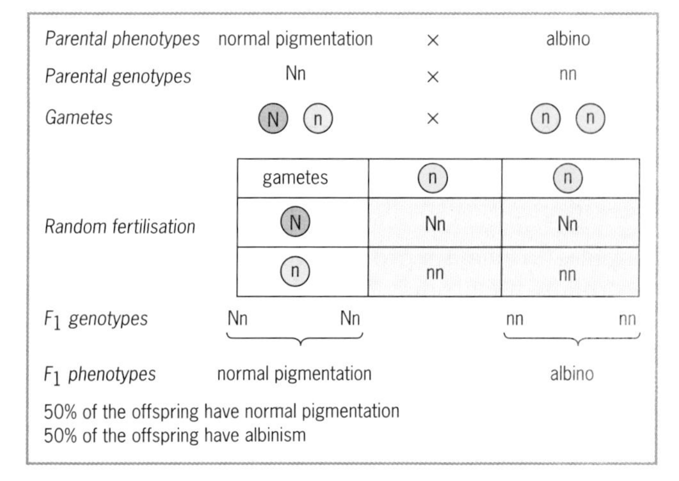

3 If both parents are **heterozygous**, i.e. carriers:

Parental phenotypes	normal pigmentation	×	normal pigmentation
Parental genotypes	Nn	×	Nn
Gametes	(N) (n)	×	(N) (n)

Random fertilisation

gametes	(N)	(n)
(N)	NN	Nn
(n)	Nn	nn

F_1 genotypes	NN	Nn	Nn	nn
F_1 phenotypes	normal pigmentation			albino

75% of the offspring have normal pigmentation
25% of the offspring have albinism

Table 16.2 *Summary of genotypic and phenotypic ratios of offspring from different crosses*

Genotype of parents	Genotypic ratio of offspring	Phenotypic ratio of offspring
homozygous dominant × homozygous dominant	100% homozygous dominant	**all** show the **dominant** trait
homozygous dominant × heterozygous	50% homozygous dominant, 50% heterozygous	**all** show the **dominant** trait
homozygous dominant × homozygous recessive	100% heterozygous	**all** show the **dominant** trait
heterozygous × heterozygous	25% homozygous dominant, 50% heterozygous, 25% homozygous recessive	**75%** show the **dominant** trait, **25%** show the **recessive** trait i.e. a **3:1** ratio
heterozygous × homozygous recessive	50% heterozygous, 50% homozygous recessive	**50%** show the **dominant** trait, **50%** show the **recessive** trait i.e. a **1:1** ratio
homozygous recessive × homozygous recessive	100% homozygous recessive	**all** show the **recessive** trait

Co-dominance

Sometimes neither allele dominates the other such that the influence of both alleles is visible in the heterozygous individual. These alleles show **co-dominance**. For example, in *Camellia*, a flowering shrub, allele $\mathbf{C^R}$ stimulates the production of **red** flowers and the allele $\mathbf{C^W}$ stimulates the production of **white** flowers. When a plant with **red** flowers, genotype $\mathbf{C^R\,C^R}$, is crossed with one with **white** flowers, genotype $\mathbf{C^W\,C^W}$, all the F_1 generation have flowers with **red** and **white patches**, genotype $\mathbf{C^R\,C^W}$. Other examples include:

- **sickle cell anaemia**
- **ABO blood groups.**

Red and white Camellia flowers

Sickle cell anaemia

The blood of a person with sickle cell anaemia contains abnormal **haemoglobin S** instead of normal **haemoglobin A**. The disease is caused by an abnormal allele. The normal allele **Hb^A** stimulates the production of normal haemoglobin A, the abnormal allele **Hb^S** stimulates the production of abnormal haemoglobin S. These alleles show co-dominance.

Table 16.3 *Possible combinations of alleles controlling haemoglobin production*

Genotype	Haemoglobin produced	Phenotype
Hb^A Hb^A	100% haemoglobin A	**Normal.**
Hb^A Hb^S	55–65% haemoglobin A 35–45% haemoglobin S	**Sickle cell trait.** Usually no symptoms. Symptoms of sickle cell anaemia may develop in very low oxygen concentrations, e.g. at high altitude or during extreme physical exercise.
Hb^S Hb^S	100% haemoglobin S	**Sickle cell anaemia.** Symptoms of sickle cell anaemia develop which include painful crises, anaemia, increased vulnerability to infections and jaundice.

Example

If both parents have **sickle cell trait**:

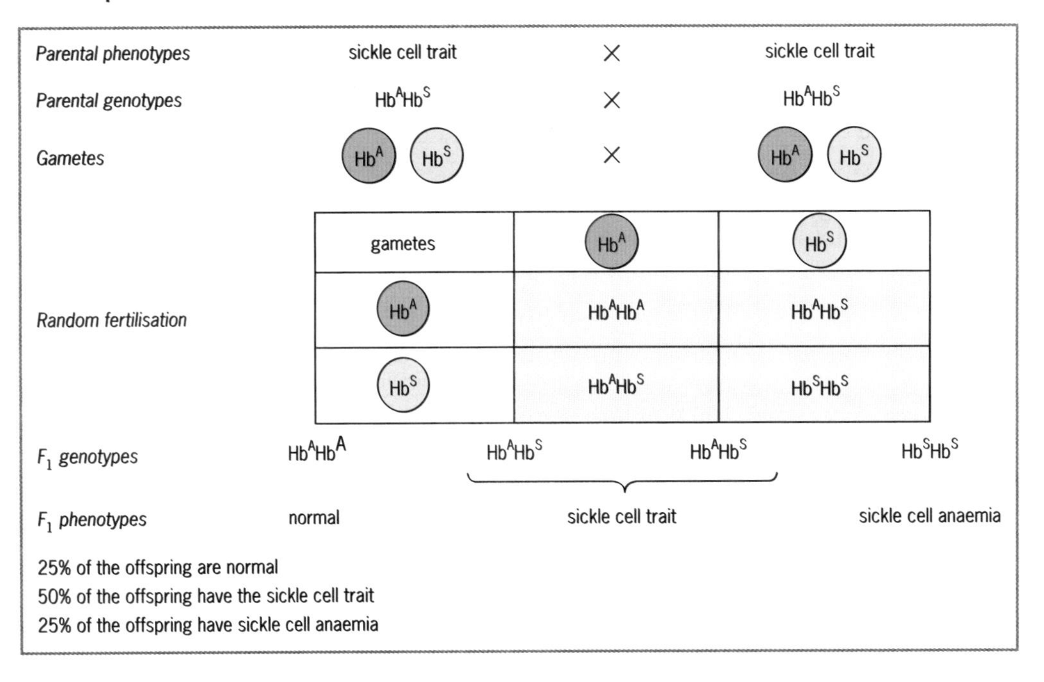

ABO blood groups

ABO blood groups are controlled by **three** alleles, **I^A**, **I^B** and **I^O**.

- **I^A** and **I^B** are both dominant to **I^O**.
- **I^A** and **I^B** are co-dominant, i.e. there is no dominance between them.

Only **two** alleles are present in any cell.

Table 16.4 *Possible combinations of alleles controlling ABO blood groups*

Genotype	Phenotype
$I^A I^A$	Blood group A
$I^A I^O$	Blood group A
$I^B I^B$	Blood group B
$I^B I^O$	Blood group B
$I^A I^B$	Blood group AB
$I^O I^O$	Blood group O

Sample question

A **heterozygous** female of **blood group A** marries a **heterozygous** male of **blood group B**. What are the chances of their first child having blood group O? Explain your answer by means of a genetic-cross diagram.

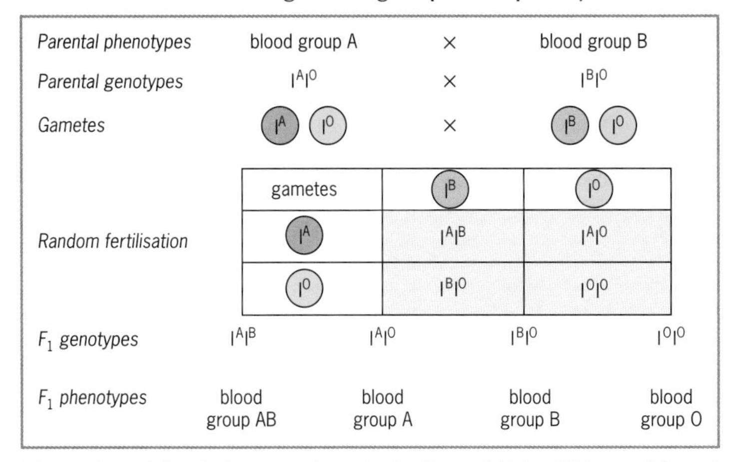

There is a **1 in 4** chance that their first child will have blood group O.

Pedigree charts

A **pedigree chart** shows how a specific trait is passed down among family members. Pedigree charts can be used to determine **genotypes**, or possible genotypes, of the individuals shown, and to predict possible genotypes and phenotypes of future offspring. This information is used by **genetic counsellors** to identify potential risks for future offspring developing a genetic disorder.

Example

A chemical substance called **PTC** tastes bitter to some people and is tasteless to others. The ability to taste PTC is controlled by a pair of alleles. The allele enabling tasting of PTC, **T**, is dominant. The non-tasting allele, **t**, is recessive.

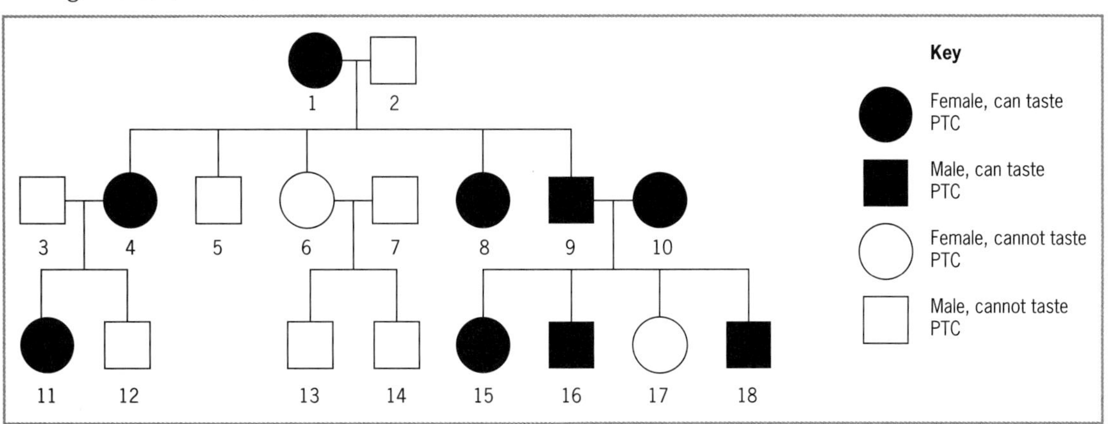

Figure 16.5 *A pedigree chart to show the inheritance of PTC tasting*

- Genotypes of individuals 2, 3, 5, 6, 7, 12, 13, 14, and 17 must be **tt** since they all have the recessive trait.
- Genotypes of individuals 1, 4, 8, 9, 10 and 11 must be **Tt** since they all had one parent who had the recessive trait or produced at least one offspring with the recessive trait.
- Genotypes of individuals 15, 16 and 18 could be **TT** or **Tt** since both parents were heterozygous.

Mechanism of sex determination

In each cell, one pair of chromosomes is composed of the **sex chromosomes**. There are two types, **X** and **Y**, and they determine the individual's gender. Genotype **XX** is **female**; genotype **XY** is **male**. Only the **male** can pass on the **Y** chromosome, consequently the **father** is the parent who determines the gender of his offspring.

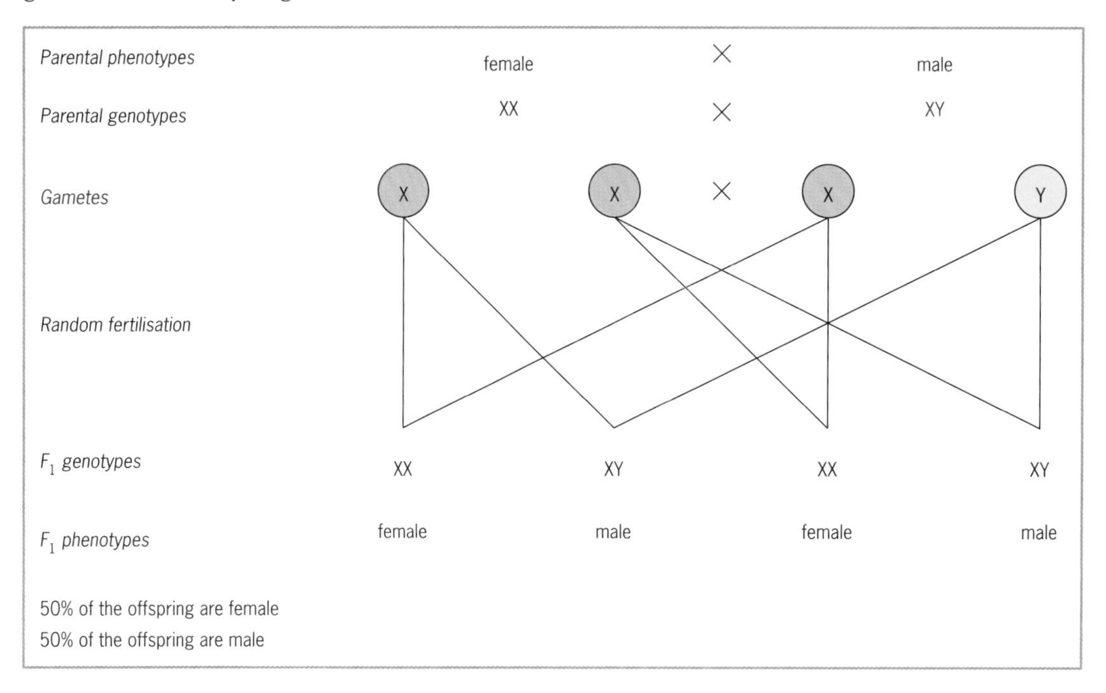

Sex-linked characteristics

Sex-linked characteristics are characteristics determined by genes carried on the sex chromosomes that have nothing to do with determining gender. These are known as **sex-linked genes**. Since chromosome X is longer than chromosome Y, it carries more genes. Males only have one X chromosome and any allele carried on this chromosome only, whether dominant or recessive, will be expressed in the phenotype.

Haemophilia

Haemophilia is a sex-linked condition where the blood fails to clot at a cut. The **dominant** allele, **H**, causes blood to clot normally; the **recessive** allele, **h**, causes haemophilia. These alleles are carried on the **X** chromosome only. Males are much more likely to have haemophilia than females; if the single X chromosome in a male carries the recessive allele he will have the condition, whereas both X chromosomes must carry the recessive allele in a female for her to have the condition.

Table 16.5 *Possible combinations of alleles controlling blood clotting*

Genotype	Phenotype
$X^H X^H$	Female, normal blood clotting
$X^H X^h$	Female, normal blood clotting (carrier)
$X^h X^h$	Female with haemophilia
$X^H Y$	Male, normal blood clotting
$X^h Y$	Male with haemophilia

Example

A cross between a **female** with **normal blood clotting** who is a **carrier**, and a **male** with **normal blood clotting**.

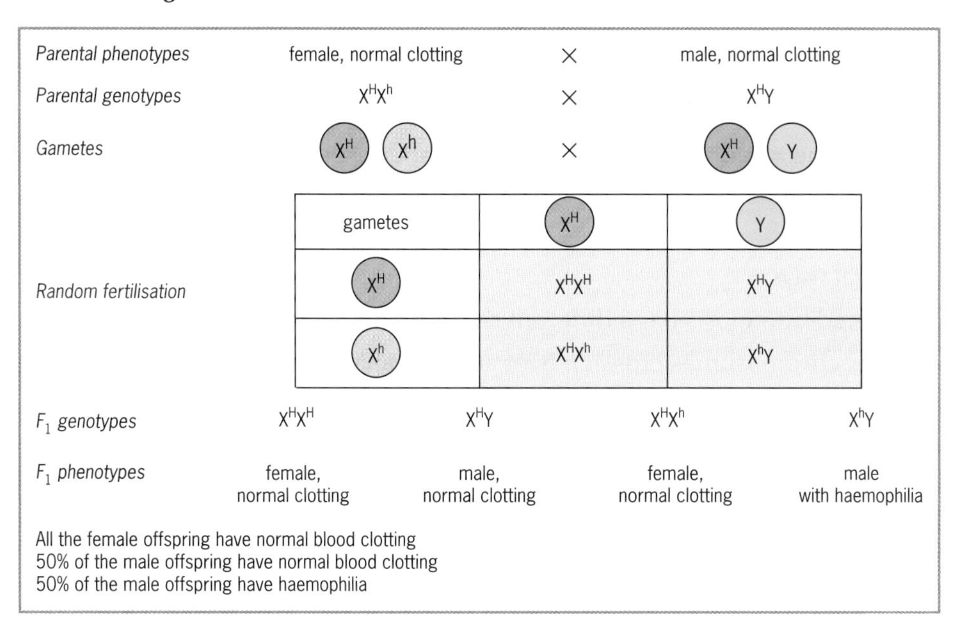

Colour blindness

Colour blindness is a sex-linked condition where the sufferer is unable to distinguish differences between certain colours. The **dominant** allele, **N**, allows normal vision and the **recessive** allele, **n**, causes colour blindness. These alleles are carried on the **X** chromosome only, so colour blindness is inherited in the same way as haemophilia.

Some important genetic terms

- ***Gene:*** *the basic unit of heredity which is composed of DNA, occupies a fixed position on a chromosome and determines a specific characteristic.*
- ***Allele:*** *either of a pair (or series) of alternative forms of a gene that occupy the same position on a particular chromosome and that control the same character.*
- ***Dominant allele:*** *the allele that, if present, produces the same phenotype whether its paired allele is identical or different.*
- ***Recessive allele:*** *the allele that only shows its effect on the phenotype if its paired allele is identical.*
- ***Dominant trait:*** *an inherited trait that results from the presence of a single dominant allele. It is seen in an individual with one or two dominant alleles.*

- ***Recessive trait:*** *an inherited trait that results from the presence of two recessive alleles. It is only seen in an individual with no dominant allele.*
- ***Co dominance:*** *neither allele dominates the other such that the influence of both alleles is visible in the heterozygous individual.*
- ***Genotype:*** *the combination of alleles present in an organism.*
- ***Phenotype:*** *the observable characteristics of an organism.*
- ***Homozygous:*** *having two identical alleles in corresponding positions on a pair of homologous chromosomes.*
- ***Heterozygous:*** *having two different alleles in corresponding positions on a pair of homologous chromosomes.*

Variation

No two living organisms are exactly alike, not even identical twins. **Variation** arises from a combination of **genetic causes** and **environmental causes**. The **phenotype** of an organism is determined by its **genotype** and the influences of its **environment**:

phenotype = genotype + environmental influences

Genetic causes of variation

Genetic variation arises in several ways:

- **Meiosis**. Every gamete produced by meiosis has a different combination of genes as a result of:
  - chromatids of homologous chromosomes crossing over and **exchanging** genes
  - chromosomes arranging themselves around the equators of the spindles in totally **random** ways.
- **Sexual reproduction**. During **fertilisation**, male and female gametes fuse in completely **random ways** to create different combinations of genes in each zygote.
- **Mutations**. A mutation is a sudden **change** in a single gene or in part of a chromosome containing several genes. Mutations cause new characteristics to suddenly develop in organisms. Mutations occurring in body cells cannot be inherited whereas mutations occurring in a gamete or zygote can be inherited. Most mutations are harmful; however, a few produce **beneficial characteristics** which provide the organism with a **selective advantage** in the struggle for survival, e.g. the peppered moth (see page 159).

Environmental causes of variation

Living organisms are constantly affected by the different factors in their **environment**. Food, drugs, physical forces, temperature and light can affect animals. Temperature, light intensity, availability of mineral salts and water all affect plants. This variation is not caused by genes and **cannot** be passed on to offspring.

The importance of variation

Variation is important because:

- It enables species to **adapt** to changing environmental conditions, improving their chances of survival.
- It provides the raw material on which **natural selection** can work, and is therefore essential for species to remain **well adapted** to their environment or to gradually **change** and **improve** by becoming better adapted to their environment.
- It makes it **less likely** that any adverse changes in environmental conditions will wipe out an entire species since some organisms may be able to adapt to the new conditions.

Types of variation

There are two basic types of variation within a species:

- **Continuous variation**

  Continuous variation is where characteristics show a **continuous gradation** from one extreme to the other without a break. Most organisms fall in the middle of the range with fewer at the two extremes, i.e. the characteristics show a **normal distribution**. Examples include height, weight, foot size, hair colour, and leaf size in plants.

  Characteristics showing continuous variation are usually controlled by **many genes** and can be affected by environmental factors.

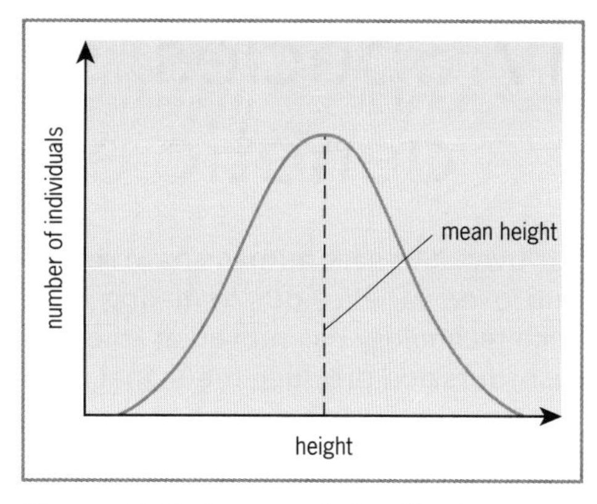

Figure 16.6 *A normal distribution curve showing height*

- **Discontinuous variation**

  Discontinuous variation is where characteristics show **clear cut differences** with no intermediates. Individuals can be divided into distinct categories, e.g. ABO blood groups, tongue rolling, and the presence or absence of horns in cattle. Characteristics showing discontinuous variation are usually controlled by a **single gene** and environmental factors have little, if any, influence on them.

Revision questions

8 Distinguish between the following pairs of terms:

a gene and allele

b genotype and phenotype

c homozygous and heterozygous.

9 PTC is a chemical that tastes bitter to some people and is tasteless to others. The ability to taste PTC is dominant. Use appropriate symbols and a genetic-cross diagram to show how a couple who can both taste PTC can produce a child who is unable to taste PTC.

10 What is co-dominance?

11 Is it possible for a female of blood group A and a male of blood group AB to have a child of blood group B? Use appropriate symbols and a genetic-cross diagram to support your answer.

12 In humans, is it the mother or father who determines the sex of their children? Explain your answer by means of a genetic diagram.

13 What are sex-linked characteristics?

14 Colour blindness is caused by an X-linked, recessive allele. Two parents with normal colour vision have a colour blind child. Use a genetic diagram to show how this is possible (X^N = normal vision; X^n = colour blindness).

15 Outline THREE ways in which genetic variation arises.

16 Give THREE reasons why it is important that living organisms show variation.

17 By reference to specific examples, distinguish between continuous and discontinuous variation.

17 Species, selection and genetic engineering

For over 200 years humans have **selected** and **crossed** plants and animals to produce new and improved varieties due to mixing genetic material. Since the structure of DNA was discovered in 1953, **gene technology** has moved at an ever increasing pace to change the traits of organisms, treat genetic disorders and produce medicinal drugs.

Species

A ***species*** *is a group of organisms of common ancestry that closely resemble each other and are normally capable of interbreeding to produce fertile offspring.*

Members of some closely related species are capable of interbreeding and producing offspring; however, their offspring are usually either **sterile** or are so **biologically weak** that they rarely produce offspring. This keeps species as **distinct groups**.

Species that can interbreed include some birds, e.g. certain species of owls, gulls, crows and ducks, and many species of plants, e.g. interbreeding of shaddock and Jamaican sweet orange created the grapefruit, and spearmint and water mint created peppermint. Mammal species that can interbreed include a donkey and a horse which produce a **mule** or a **hinney**, and a lion and a tiger which produce a **liger** or a **tigon**.

The formation of new species – speciation

As long as organisms from different groups within a species can interbreed and genes can flow between them, the groups remain members of the same species. If groups become **separated** or **isolated**, the flow of genes between them stops. **Genetic differences** gradually develop and a point is reached where members of the groups can no longer successfully interbreed. They become **separate species**, each with its own pool of genes.

- **Speciation by geographical separation**

  This occurs when a **physical barrier** prevents two groups of organisms of the same species from meeting and interbreeding. Such barriers include mountain ranges, deserts, oceans, rivers or even streams.

- **Speciation by ecological and behavioural separation**

  Speciation can occur when two groups of organisms of the same species inhabit the same region but they become adapted to different **habitats** in that region, which reduces gene flow.

  Speciation can also occur when animals exhibit elaborate **courtship behaviours** before mating, which may be stimulated by the colour, markings, calls or actions of the opposite sex. If small differences occur in any of these stimuli it can prevent mating, which prevents gene flow.

Extinction of species

Over time species can also become **extinct**, i.e. they no longer exist. Habitat loss, disease, predation by introduced species, competition with introduced species or overexploitation by humans, e.g. overfishing or overhunting, can all lead to extinction of species. For example, the **Caribbean monk seal** has become extinct due to it being overhunted for its fur, meat and oil.

The role of natural selection in biological evolution

Natural selection is the process by which populations change over time or **evolve**, so that they remain well adapted to their environment. Charles Darwin was the first person to put forward the idea of evolution by natural selection in 1859 in his book *On the Origin of Species*. The **theory of natural selection** is based on the following:

- Most organisms produce **more offspring** than are needed to replace them, yet the numbers of individuals in populations remain relatively constant. In nature there must, therefore, be a constant **struggle for survival**.
- All organisms show **variation** and much of this can be inherited. Those organisms possessing variations that make them **well adapted** to their environment are most likely to survive in the struggle, i.e. there is **survival of the fittest**.
- Since the well adapted organisms are the most likely to survive, they are the ones most likely to **reproduce**, thereby passing on their advantageous characteristics to their offspring. Species, therefore, remain well adapted to their environment or they gradually **change** and **improve** by becoming even better adapted.

Natural selection **preserves useful adaptations** since the genes that produce advantageous characteristics are passed on to offspring **more frequently** than the genes that produce less advantageous characteristics. It is the mechanism by which populations retain the genes that make them well adapted to their environments.

Genetic variation, especially that resulting from beneficial mutations, is the **raw material** for natural selection.

Evidence for natural selection

Natural selection in action can be seen in the following examples:

- **The peppered moth**

  The **peppered moth** lives in Britain and is eaten by birds. Before the Industrial Revolution, the moths were black and white speckled and were well camouflaged against the pale lichen-covered tree trunks on which they rested.

  During the Industrial Revolution, a **melanic** (all black) variety appeared in the industrial area around Manchester. This melanic variety arose as a result of a dominant mutation and was well camouflaged against the tree trunks which were blackened with soot. This gave the melanic variety a **selective advantage** in industrial areas and, over time, it became far more numerous in these areas than the speckled variety.

Peppered moth

- **Antibiotic and pesticide resistance**

  In natural populations of bacteria and various pests, e.g. insects, fungi and weeds, a few individuals may carry genes that make them **resistant** to antibiotics or various pesticides, e.g. insecticides, fungicides and herbicides. These genes arise from mutations. When exposed to antibiotics or pesticides, these resistant organisms have a **selective advantage**; they are more likely to survive and reproduce than non-resistant organisms, passing on their resistance to their offspring. This is causing increasing numbers of resistant organisms to appear within populations.

- **Galapagos finches**

  The Galapagos Islands in the Pacific Ocean have at least 13 different species of finches which are possibly all descendants of a single South American species that colonised the islands from the mainland.

Two of Darwin's finches

The main difference between species is in the shape and size of their beaks. As a result of **natural selection**, their beaks have become highly adapted to the different types of food present on the various islands, e.g. seeds, insects, nectar or fruits.

- **Caribbean lizards**

  Anole lizards are thought to have colonised the islands of the Caribbean from Central and South America. Through **natural selection**, lizards stranded on the four larger islands of Cuba, Hispaniola, Jamaica and Puerto Rico independently evolved into different species with similar characteristics that enabled them to fit similar ecological niches on each island, e.g. twig anoles developed long, slender bodies and tails and short legs; trunk ground anoles developed long, muscular legs, and canopy anoles developed large toe pads. Today, the different species have **equivalent species** with similar body types on each island.

A twig anole

Artificial selection

Artificial selection involves humans selecting and breeding organisms showing **desirable characteristics**. As a result, new breeds, strains or varieties of plants and animals are produced with characteristics to suit human needs. Undesirable characteristics are 'bred out'. Artificial selection produces new varieties of organisms in a **shorter time** than natural selection. It has the disadvantage that it **reduces variation** in populations making them more vulnerable if environmental conditions change.

- **Inbreeding** involves breeding **closely related** individuals showing desirable characteristics. It is usually used to improve one particular trait. Continued inbreeding reduces the gene pool which increases the frequency of **undesirable genes** and reduces the overall fitness of the organisms. After several generations of inbreeding, outbreeding must take place to introduce new genes into a population.
- **Outbreeding** involves breeding individuals from **genetically distinct** populations showing desirable characteristics. Offspring produced are called **hybrids** and usually show characteristics that are superior to both parents. This is known as **hybrid vigour**.

Artificial selection is used extensively in agriculture to produce crop plants and farm animals with:

- **Increased yields**, e.g. cattle that produce more milk or meat, chickens that lay more or larger eggs, sugar cane that produces more sucrose and cereal crops that produce more grain.
- **Increased quality** of product, e.g. meat with less fat, and cereals and ground provisions with a higher protein content.
- **Faster growth rates**.
- **Increased number of offspring**.
- **Shorter time to reach maturity** so that more generations are produced per year.
- **Increased resistance** to pests and disease. This reduces product loss and the need for pesticides.
- **Increased suitability to the environment**.

Artificial selection in action in the Caribbean

- **Jamaica Hope**, a breed of dairy cattle, was developed in Jamaica by breeding Jersey, Zebu and Holstein cattle. The breed is heat tolerant, has a high resistance to ticks and tick borne diseases, and produces a high yield of milk, even when grazing on the poor pasturelands of the Caribbean.
- **Sugar cane** has been bred to produce varieties with a high sucrose content, increased resistance to disease and insect pests, greater suitability to its environment and improved ratooning ability.

Genetic engineering

Genetic engineering involves changing the traits of one organism by inserting genetic material from a different organism into its DNA. The organism receiving the genetic material is called a **transgenic organism** or **genetically modified organism (GMO)**.

Genetic engineering is used to:

- Protect agricultural crops against environmental threats, e.g. pathogens, pests, herbicides and low temperatures.
- Modify the quality of a product, e.g. increasing nutritional value.
- Make organisms produce materials that they do not usually produce, e.g. vaccines and drugs.
- Improve yields, e.g. increasing size or growth rate, or making organisms more hardy.

Genetic engineering and food production

Genetic engineering is used to improve food production.

Examples

- **Golden rice**

  By inserting two genes into rice plants, one from maize and one from a soil bacterium, the endosperm of the rice grains is stimulated to produce **beta-carotene** which the body converts to vitamin A. Golden rice should help fight vitamin A deficiency which is a leading cause of blindness, and often death, of children in many underdeveloped countries.

Golden rice

- **Roundup resistant crops**

  By inserting a gene from a soil bacterium into certain crop plants, e.g. soya bean, corn and canola, the plants become **resistant** to the herbicide called 'Roundup'. The herbicide can be sprayed on the crops to destroy weeds, but not harm the crops.

- **Bt corn**

  By inserting a gene from a soil bacterium into corn plants, the plants are stimulated to produce a chemical that is toxic to corn-boring caterpillars. This makes the corn plants **resistant** to the caterpillars.

- **Bovine somatotrophin (BST) hormone**

  By transferring the gene that controls the production of BST hormone from cattle into bacteria, the bacteria produce the hormone which is then injected into cattle to increase milk and meat production.

- **Chymosin (rennin)**

  By transferring the gene that controls the production of chymosin from calf stomach cells into bacteria or fungi, the micro-organisms produce chymosin which is used in **cheese** production. This has considerably increased the production of cheese worldwide.

Genetic engineering and medical treatment

Genetic engineering is used to produce many drugs used in medical treatment.

Examples

- **Insulin**

  By transferring the gene that controls insulin production in humans into bacteria, the bacteria produce insulin which is used to treat **diabetes**.

- **Human growth hormone (HGH)**

  By transferring the gene controlling the production of HGH into bacteria, the bacteria produce the hormone which is used to treat **growth disorders** in children.

- **Hepatitis B vaccine**

  By transferring the gene controlling the production of hepatitis B antigens in the hepatitis B virus into yeast, the yeast produces the **antigens** which are used as a **vaccine**.

Other **drugs** produced by genetic engineering include:

- **Blood clotting drugs** for people with haemophilia.
- **Follicle stimulating hormone (FSH)** used to stimulate the ovaries to produce mature ova in women that are infertile.
- **Interferons** used to treat viral infections and certain cancers.
- **Anticoagulants** used to prevent the development of life-threatening blood clots in heart patients.
- **Human papilloma virus vaccine**.

Possible advantages of genetic engineering

- **Yields** can be **increased** by genetic engineering which should increase the world food supply and reduce food shortages.
- The **nutritional value** of foods can be increased by genetic engineering which should reduce deficiency diseases worldwide.
- The need for **chemical pesticides** that harm the environment can be reduced by genetically engineering crops to be resistant to pests.
- **Vaccines** produced by genetic engineering are generally **safer** than vaccines containing live and weakened, or dead pathogens.
- **Larger quantities** of drugs in a **safer** and **purer** form can be produced than were previously produced from animal sources resulting in more people worldwide having ready access to safe, life-saving drugs.
- It overcomes **ethical concerns** of obtaining certain drugs from animals, e.g. insulin used to be obtained from pigs and cows.

Possible disadvantages of genetic engineering

- Plants genetically engineered to be toxic to a pest may also be toxic to **useful organisms**, e.g. insects that bring about pollination. This could negatively affect wild plants and reduce reproduction in crops, reducing food production.
- Plants genetically engineered to be resistant to pests and herbicides could create **unpredictable environmental issues**, e.g. they could lead to the development of pesticide-resistant insects or they could interbreed with closely related wild plants and create herbicide-resistant superweeds.

- Once a genetically modified organism is released into the environment, it cannot be **contained** or **recalled.** Any negative effects are irreversible.
- The number of **allergens** in foods could be increased by transferring genes causing allergic reactions between species.
- As yet **unknown health risks** may occur as a result of eating genetically modified plants and animals.
- Large companies with funds and technology to develop genetically modified organisms could make **large profits** at the expense of smaller companies and poorer nations.
- Future steps in genetic engineering might allow the genetic make-up of higher organisms, including humans, to be altered, e.g. to produce 'designer babies'. Difficult **moral** and **ethical issues** then arise, e.g. how far should we go in changing our own genes and those of other animals?

Other applications of gene technology

- **DNA testing or DNA fingerprinting**

  DNA testing involves analysing specific regions of DNA taken from cells of individuals, scenes of accidents or crime scenes. It is used:
  - To determine if two DNA samples are from the same person thereby helping to **solve crimes.**
  - To determine the **paternity** and, in some cases, the maternity of a child.
  - To **identify** a body.
  - To detect **genetic disorders** or **diseases** before birth or early in life so treatment can begin at an early age.
  - To help **genetic counsellors** predict the likelihood that a child who is born to parents who have a genetic disease, or are carriers of a genetic disease, will suffer from the disease.
  - To identify **family relationships** thereby reuniting families.
  - To determine **ancestral lines** and create family trees.

- **Gene therapy**

  Gene therapy is an experimental technique that involves **altering genes** inside body cells to cure a disease or help the body fight a disease. It is currently being tested for use in various ways:
  - By inserting a **functional gene** into cells to replace a defective gene that causes a disease.
  - By **inactivating** or 'turning off' a defective gene that causes a disease.
  - By introducing a gene into cells to help the body's immune system to **fight** a disease.

- **Captive breeding programmes**

  Captive breeding involves breeding and raising animals in human controlled environments, e.g. zoos, aquaria and wildlife reserves. The aim is to prevent the extinction of endangered species, conserve species that may not survive well in the wild, reintroduce animals back into the wild and preserve biodiversity. **DNA profiling** is used in these programmes:
  - To assess the **genetic diversity** of organisms to be bred thereby preventing breeding organisms that are too genetically similar, i.e. inbreeding (see page 160).
  - To help prevent the **loss** of genetic diversity within offspring and future generations of offspring produced by the breeding programmes.

Revision questions

1. Define the term 'species'.
2. Identify TWO ways in which new species may develop.
3. Explain how natural selection plays a role in biological evolution.
4. There are several pieces of evidence in existence today that support the theory of natural selection. Discuss TWO of these.
5. Humans are able to apply the principles of natural selection in agriculture. Using ONE plant and ONE animal as examples, explain how this is being done in the Caribbean.
6. What is genetic engineering?
7. Describe TWO ways in which genetic engineering is being used to improve food production and TWO ways in which it is being used in medical treatment.
8. Discuss THREE possible advantages and THREE possible disadvantages of genetic engineering.
9. Outline THREE uses of DNA testing.

Exam-style questions – Chapters 16 to 17

Structured questions

1 **a)** Figure 1 shows a dividing cell.

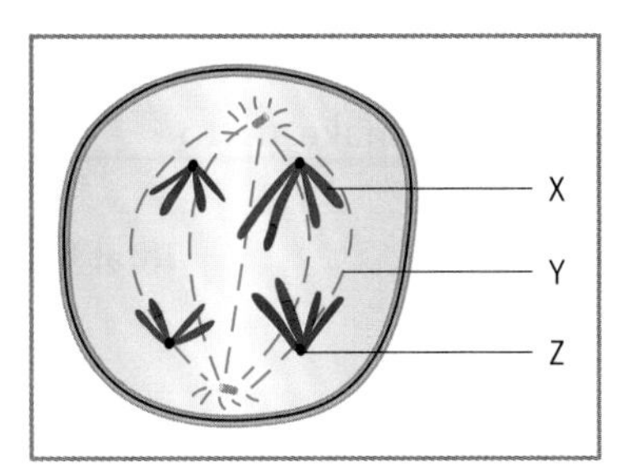

Figure 1 *A dividing cell*

- **i)** Name the structures labelled X, Y and Z. **(3 marks)**
- **ii)** Identify the type of cell division shown in Figure 1. **(1 mark)**
- **iii)** Give TWO reasons for your answer to **ii**). **(2 marks)**
- **iv)** Name ONE place in the human body where they type of cell division shown in Figure 1 would occur. **(1 mark)**
- **iv)** How many chromosomes would each daughter cell possess when the cell shown in Figure 1 has finished dividing? **(1 mark)**
- **v)** Give TWO ways in which the cell division shown in Figure 1 differs from the cell division occurring in a growing embryo. **(2 marks)**

b)
- **i)** What is cloning? **(2 marks)**
- **ii)** Despite the fact that sugar cane can reproduce sexually, most sugar cane is grown from stem cuttings. Suggest TWO advantages of using this method. **(2 marks)**
- **iii)** Suggest ONE disadvantage of cloning in plants. **(1 mark)**

Total 15 marks

2 Albinism is a condition in which the external pigmentation fails to develop. It is caused by a recessive allele and is not sex-linked. The family tree in Figure 2 below shows the inheritance of albinism in a family.

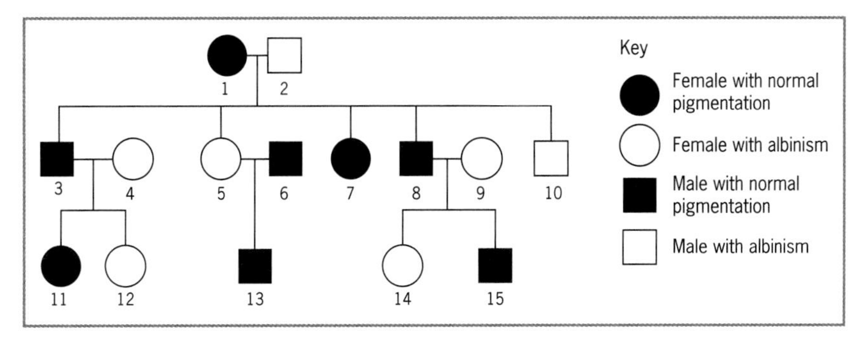

Figure 2 *The inheritance of albinism in a family*

a) i) Using appropriate symbols, give the genotypes of persons 1 and 2. **(2 marks)**

ii) If persons 11 and 15 were to marry, what is the chance that any of their offspring would have albinism? Using a genetic cross diagram, explain your answer. **(3 marks)**

iii) If the albinism was sex-liked, give the possible phenotypes of the children of persons 5 and 6. Use a genetic-cross diagram to explain your answer. **(4 marks)**

b) i) What type of variation is shown in Figure 2? **(1 mark)**

ii) Identify TWO other human characteristics that show the same type of variation as that shown in Figure 2. **(2 marks)**

iii) Give TWO reasons why variation among living organisms is important. **(2 marks)**

iv) How is it possible for organisms that have the identical genetic make-up to show variation? **(1 mark)**

Total 15 marks

Extended response questions

3 a) Distinguish between the following pairs of terms:

i) genotype and phenotype

ii) dominant trait and recessive trait. **(4 marks)**

b) In pea plants, resistance to fungal disease is a recessive trait. Use appropriate symbols and a genetic diagram to work out the possible phenotypes from a cross between a non-resistant, heterozygous plant and a resistant plant. **(4 marks)**

c) Genetic engineering is used to change the traits of organisms and it may be used to cure genetic diseases in the future.

i) Outline TWO ways genetic engineering is currently being used to improve medical treatment and discuss how it may be used in the future to cure genetic diseases. **(5 marks)**

ii) Give TWO concerns that people might have about the use of genetic engineering to change the traits of organisms. **(2 marks)**

Total 15 marks

4 a) i) What is a species and what maintains species as distinct groups? **(3 marks)**

ii) Explain TWO ways in which new species can form. **(4 marks)**

b) Increasing numbers of bacteria are becoming resistant to commonly used antibiotics. Use the theory of natural selection to help explain how this situation arises. **(4 marks)**

c) By referring to specific examples other than resistance, distinguish between artificial selection and natural selection. **(4 marks)**

Total 15 marks

Index